Accurate Case Outcome Modeling

Arch G. Woodside
Editor

Accurate Case Outcome Modeling

Entrepreneur Policy, Management, and Strategy Applications

Springer

Editor
Arch G. Woodside
Yonsei University, Yonsei Frontier Lab
Seoul, Republic of Korea

ISBN 978-3-030-26817-6 ISBN 978-3-030-26818-3 (eBook)
https://doi.org/10.1007/978-3-030-26818-3

This Springer imprint is published by the registered company Springer Nature Switzerland AG
The registered company address is: Gewerbestrasse 11, 6330 Cham, Switzerland

This book is dedicated to anomalies, contrarians, exceptions, outliers, and/or mavericks. These cases are subsegments of consumers, decisions, executives, firms, managers, nations, organizations, and other things and events. "An anomaly is a fact that doesn't fit received wisdom ... an anomaly marks an opportunity to learn something very valuable. In science, anomalies are the frontier, where the action is" (Rumelt, 2011, pp. 247–248).

Reference

Rumelt, R. P. (2011). *Good strategy/bad strategy. Profile books*. London, UK: Profile Books.

Preface: Embracing Case Outcome Modeling in Theory Construction and Empirical Research in the Social Sciences and Business Subdisciplines

The following observations represent the unifying theoretical and methodological tenets observable in all the chapters in this volume. First, construction and confirmation of *accurate case outcome modeling* (ACOM) are achievable in the behavioral science and business subdisciplines—even though at the start of the 2020s, variable-directional hypothesis construction with null hypothesis significance testing (NHST) remains in dominant logic in these subdisciplines. Hubbard's (2015, p. 191–192) insightful observation is now applicable in practice (not only in principle) in these subdisciplines: "In principle, there is no reason why theories in the management and social sciences cannot yield precise (or interval) predictions… however, this line of thinking flies in the face of conventional wisdom that theories in these areas are unable to specify point predictions." All the chapters in ACOM construct case outcome theories and include empirical analysis of outcomes using *somewhat precise outcome testing* (SPOT) (e.g., screening conditions for identifying outcome accurately and using odds ratios).

Thus, the ACOM chapters illustrate how to avoid the bad science practices of constructing theory using variable relationship-directional hypothesis and testing theory using NHST—what Hubbard (2015) refers to in the main title of his book as *Corrupt Research*—and the ACOM chapters provide in-depth examples of using good science practice. The absence of research examples, complete with findings for "statistical sameness testing" (Hubbard, 2015, p. 4), is the principal shortcoming in Hubbard (2015). Following an introductory chapter in the present volume fills this gap. Hubbard's (2015) expression, "statistical sameness testing" (SST), is examining the frequency of achieving the same outcome occurring among cases for a specific model of simple or a complex antecedent conditions. *Somewhat precise outcome testing* (SPOT) is synonymous with SST. In a personal communication, Hubbard (2016) expressed surprise that his book was receiving scant attention. However, eliminating the pervasive use of research practices that Hubbard (2015) identifies correctly to be corrupt takes time and knowledge by many researchers that, indeed, the practices are corrupt—and equally important—knowledge and skills in using good science practices.

ACOM, this book, provides theory construction and testing applications in several subdisciplines of business and social sciences to illustrate how to move away from symmetric theory construction and statistical testing tools—procedures that are bad science practice as Hubbard (2015) and others (Armstrong, 1970, 2012; Fiss, 2007, 2011; McCloskey, 2002, Trafimow, 2014; Trafimow & Marks, 2015; Wasserstein, 2016; Wasserstein & Lazar, 2016; Woodside 2013; 2017; Woodside et al. 2015; Ziliak & McCloskey, 2008, 2009) advocate replacing. For example, the principal conclusion in the American Statistical Association report (Wasserstein & Lazar, 2016, p. 131) covering bad practice in statistical analysis is that, "By itself, a *p*-value does not provide a good measure of evidence regarding a model or hypothesis." However, most researchers appear to be unaware or wish to ignore the need for replacing symmetric modeling (e.g., multiple regression analysis) and NHST and using good science practices (e.g., asymmetric case outcome modeling and SPOT). The continuing focus in teaching only symmetric statistical modeling tools almost exclusively in undergraduate and graduate degree programs in business and social science disciplines appears to be the primary causal condition of the current state of bad research modeling practices—as the ASA (Wasserstein & Lazar, 2016, p. 129) explains, "We teach it because it's what we do; we do it because it's what we teach." Thus, adding good science practices on what to do while continuing to teach bad practices as what not to do is a giant step forward for improving research practice generally.

ACOM includes six chapters. Chapter 1 is a foundation description of symmetric variable directional-relationship theory construction and null hypothesis significance testing versus asymmetric case outcome theory construction and somewhat precise outcome testing. Chapters 2, 3, 4, 5, and 6 are example applications of constructing and testing asymmetric case outcome models for project management assessment (Chap. 2), assessing strategy management success/failure outcomes (Chap. 3), worker choices of work environments (Chap. 4), customers' assessments of traditional local markets (Chap. 5), and national cultures' consequences on need motivations, entrepreneurship, innovation, and nations' ethical behavior and quality of life (Chap. 6).

Building accurate algorithms, filters, "computing with words" (Zadeh, 1996), and "ethnographic decision true models" (Gladwin, 1989) are all examples of point or interval predictive models of outcomes. Such theory construction and testing embrace the inclusion contextual and process configurational conditions. While symmetric modeling with null hypothesis significance tests (NHST) represents shallow testing of shallow directional theories, ACOMs are asymmetric "deep dives" into understanding, describing, and predicting point outcomes or interval outcomes (e.g., top or bottom quintile cases by performance). Most symmetric theory and data analysis studies focus on attempts to reject the null hypothesis that one or more independent variables (W or X) relate systematically to a dependent variable (Y), that is, increasing in X associates with increases in Y. Asymmetric theory and data analysis build on the complexity theory perspective that high X associates with high Y, high X associates with low Y, low X associates with high Y, and low X associates with low Y—which of these four outcomes occur depends on additional process and contextual conditions occur for each XY combination. Given that anomalies almost always occur in a set of data whereby cases are observable in four XY corners, just testing for a symmetric positive or negative XY directional relationship is inadequate—shallow research. For learning to perform deep dives in constructing and testing models, read on.

Seoul, Republic of Korea Arch G. Woodside

References

Armstrong, J. S. (1970). How to avoid exploratory research. *Journal of Advertising Research, 10*, 27–30.

Armstrong, J. S. (2012). Illusions in regression analysis. *International Journal of Forecasting, 28*, 689–694.

Fiss, P. C. (2007). A set-theoretic approach to organizational configurations. *Academy of Management Review, 32*, 1180–1198.

Fiss, P. C. (2011). Building better casual theories: A fuzzy set approach to typologies in organizational research. Academy. The ASA's statement on p-values: Context, process, and purpose. *The American Statistician of Management Journal, 54*, 393–420.

Gladwin, C. (1989). *Ethnographic decision tree modeling*. Newbury Park, CA: SAGE.

Hubbard, R. (2015). Corrupt research: The case for reconceptualizing empirical management and social science. Thousand Oaks, CA: Sage.

Hubbard, R. (2016). Personal communication with Arch G. Woodside, March 2, 2016 by telephone.

Trafimow, D. (2014). Editorial. *Basic and Applied Social Psychology, 36*, 1–2.

Trafimow, D., & Marks, M. (2015). Editorial. *Basic and Applied Social Psychology, 37*, 1–2.

Wasserstein, R. I. (Ed.). (2016). ASA statement on statistical significance and P-values (70, pp. 131–133). The American Statistician.

Wasserstein, R. L., & Lazar, N. A. (2016). The ASA's statement on p-values: Context, process, and purpose. *The American Statistician, 70*, 129–133.

Woodside, A. G. (2013). Moving beyond multiple regression analysis to algorithms: Calling for adoption of a paradigm shift from symmetric to symmetric thinking in data analysis and crafting theory. *Journal of Business Research, 66*, 463–472.

Woodside, A. G. (2017). Releasing the death-grip of null hypothesis statistical testing (p< .05): Applying complexity theory and somewhat precise outcome testing (SPOT). *Journal of Global Scholars of Marketing Science, 27*, 1–15.

Woodside, A.G., Prentice, C., Larsen, A. (2015). Revisiting problem gamblers' harsh gaze on casino services: applying complexity theory to identify exceptional customers. *Psychology & Marketing, 32*(1), 65–77.

Zadeh, L. A. (1996). Fuzzy logic: Computing with words. *IEEE Transactions on Fuzzy Systems, 4*, 103–111.

Ziliak, S. T., & McCloskey, D. N. (2008). The cult of statistical significance: How the standard error costs us jobs, justice and lives. University of Michigan Press. Ann Arbor MI.

Ziliak, S. T., & McCloskey, D. N. (2009). The cult of statistical significance. Section on statistical education – Joint statistical meetings.

Acknowledgments

Acknowledgment is given to the brilliant works of the following troublemakers in contributing to the destruction of bad science practices—the then-and-now pervasive logic of variable directional-relationship hypothesis and NHST. Reading these rabble-rousing contributions compelled ACOM's conception, gestation, and birth: Armstrong (2012), Gigerenzer, Todd, and the ABC Research Group (2000), Gladwin (1989), Howard and Morgenroth (1968), Hubbard (2015), McClelland (1998), Ragin (2008), Simon (1978), Zadeh (1965), Ziliak and McCloskey (2008).

References

Armstrong, J. S. (2012). Illusions in regression analysis. *International Journal of Forecasting, 28*(3), 689–694.

Gigerenzer, G., Todd, P. M., & the ABC Research Group. (2000). *Simple heuristics that make us smart*. Oxford, UK: Oxford University Press.

Gladwin, C. (1989). *Ethnographic decision tree modeling*. Newbury Park, CA: SAGE.

Howard, J. A., & Morgenroth, W. M. (1968). Information processing model of executive decision. *Management Science, 14*(7), 416–428.

Hubbard, R. (2015). *Corrupt Research: The case for reconceptualizing empirical management and social science*. Los Angeles, CA: Sage.

McClelland, D. C. (1998). Identifying competencies with behavioral-event inter-views. *Psychological Science, 9*, 331–339.

Ragin, C. C. (2008). *Redesigning social inquiry*. Chicago: University of Chicago Press.

Simon, H. A. (1978). *Rational decision-making in business organizations*. Prize Lecture. NobelPrize.org. Nobel Media AB 2018. Available at https://www.nobelprize.org/uploads/2018/06/simon-lecture.pdf. Accessed on 18 October 2018.

Zadeh, L. (1965). Fuzzy sets. *Information and Control, 8*, 338–335.

Ziliak, S. T., & McCloskey, D. N. (2008). *The cult of statistical significance: How the standard error costs us jobs, justice and lives*. Ann Arbor, MI: University of Michigan Press.

Contents

Contributors

Olajumoke A. Awe Coastal Carolina University, Wall College of Business of Administration, Conway, SC, USA

Graham Ferguson Curtin University, School of Marketing, Bentley Campus, Perth, WA, Australia

Lars Isaksson Queensland University of Technology, Building Z, School of Advertising, Marketing, and Public Relations, Brisbane, QLD, Australia

Jaesuk Jung Yonsei University, College of Human Ecology, Seoul, Republic of Korea

Eunju Ko Yonsei University, College of Human Ecology, Seoul, Republic of Korea

Carol M. Megehee Coastal Carolina University, Wall College of Business Administration, Conway, SC, USA

Gábor Nagy INSEEC Paris Business School, Department of Marketing, Paris, France

Sridhar Nerur University of Texas, Department of Information Systems and Operations Management, Arlington, TX, USA

Edmund Prater University of Texas, Department of Information Systems and Operations Management, Arlington, TX, USA

April J. Spivack Coastal Carolina University, Wall College of Business Administration, Conway, SC, USA

Arch G. Woodside Yonsei University, Yonsei Frontier Lab, Seoul, Republic of Korea

About the Editor

Arch G. Woodside is visiting research professor, Coastal Carolina University, USA; distinguished university professor, Yonsei University, Yonsei Frontier Lab, Seoul, ROK; and honorary professor of marketing, Curtin University, Australia. He is the editor in chief of the *Journal of Global Scholars of Marketing Science* and editor of *Advances in Culture, Tourism, and Hospitality Research*. Research articles (co)authored by Dr. Woodside appear in 55 hospitality, management, marketing, psychology, and tourism journals. His book publications include *Brand Choice Revealing Customers' Unconscious-Automatic and Strategic Thinking Processes* (with Randolph J. Trappey III, Palgrave 2004); *The Complexity Turn* (ed., Springer 2017); and *Incompetency and Competency Training: Improving Executive Skills in Sensemaking, Framing Issues, and Making Choices* (with Rouxelle de Villiers and Roger Marshall, Springer 2016).

Chapter 1
Matching Case Identification Hypotheses and Case-Level Data Analysis

Arch G. Woodside

Abstract The traditional and still dominant logic among nearly all empirical positivist researchers in schools of management is to write symmetric (two directional) variable hypotheses (SVH), even though the same researchers formulate their behavioral theories at the case (typology) identification level. *The behavioral theory of the firm* (Cyert and March, A behavioral theory of the firm. Prentice-Hall, Englewood Cliffs, 1963), *the theory of buyer behavior* (Howard and Sheth, The theory of Buyer behavior. Wiley, New York, 1969)), and Miles and Snow's (Organizational strategy, structure, and process. McGraw Hill, New York, 1978) typologies of organizations' strategy configurations (e.g., "prospectors, analyzers, and defenders") are iconic examples of formulating theory at the case identification level. When testing such theories, most researchers automatically, nonconsciously, switch from building theory of beliefs, attitudes, and behavior at the case identification level to empirically testing of two-directional relationships and additive net-effect influences of variables. Formulating theory focusing on creating case identification hypotheses (CIH) to describe, explain, and predict behavior and then empirically testing at SVH is a mismatch and results in shallow data analysis and frequently inaccurate contributions to theory. This chapter describes the mismatch and resulting unattractive outcomes as well as the pervasive practice of examining only fit validity in empirical studies using symmetric tests. The chapter reviews studies in the literature showing how matching both case-based theory and empirical positivist research of CIH is possible and produces findings that advance useful theory and critical thinking by executives and researchers.

Keywords Behavior · Case · Empiricism · Management · Theory · Variable

A. G. Woodside (✉)
Yonsei University, Yonsei Frontier Lab, Seoul, Republic of Korea
e-mail: arch.woodside@bc.edu

A. G. Woodside (ed.), *Accurate Case Outcome Modeling*,
https://doi.org/10.1007/978-3-030-26818-3_1

1 Introduction

Most empirical research in behavioral science and business research journals mismatch theory and data analysis. Most researchers develop theory from the perspective of the individual firm or consumer but formulate their hypotheses from the perspective of the net-effect influence of individual variables on a dependent variable. The researchers' shift in thinking is seemingly subtle and likely an unconscious one. Their prior training in data analyses focuses on how to do symmetric tests such as analysis of variance, multiple regression analysis (MRA), and structural equation modeling (SEM); they apply this training to extract information from their data without recognizing the transformation in their focus from case (e.g., firm, consumer, organization, nation) identification into a symmetric variable hypotheses (SVH) perspective. Symmetric tests of variable hypotheses examine two-way directional hypotheses such as high versus low X (independent) scores associated with high versus low Y scores. For example, Hofstede's (2001) cultural value theory proposes that each nation is a complex whole of a combination of distinct cultural values (e.g., collectivism/individualism, masculinity, uncertainty avoidance, and power distance); a vast number of studies examine Hofstede's theory in many different contexts, but almost all of these studies examine the net effects of each cultural value using symmetric tests. Almost none of these studies examine the influence of culture values from the perspective of culture as complex wholes (i.e., recipes), including the studies by Hofstede and colleagues; for exceptions, see Hsu, Woodside, and Marshall (2013); Woodside, Hsu, & Marshall (2011).

The majority of studies in scholarly behavioral science and business-related journals on many different topics exhibit this theory-analysis mismatch. Most of the resulting published articles report low levels of explained variance (R^2) in their dependent variables in the findings section and struggle in their discussion sections to show how net effects of individual variables are relevant for complex wholes of the firm, person, or organization. While most of the relevant literature fails to acknowledge this mismatch specifically—except for McClelland (1998) and Fiss (2007)—a few researchers (Armstrong 2012; Bass et al. 1968; Gigerenzer and Brighton 2009; Ordanini et al. 2014; Ragin 2008) do describe problems with the still current dominant logic of reporting findings using symmetric tests; these authors offer helpful solutions to overcome these problems. A bit of headway is now occurring in doing what McClelland advocated in the 1990s—taking steps to overcome the limitations of using regression analysis (symmetric tests) and the mismatching of theory and data analysis. Possibly a tipping point is appearing in the literature due to the subsequent work of Meier and Donzé (2012), Fiss (2007, 2011), Fiss, Marx, and Cambré (2013), Ordanini et al. (2014), Ragin (2008), Woodside, Hsu, and Marshall (2011), Woodside (2013), and the studies by researchers who are members of COMPASSS.ORG.

This chapter describes the mismatch between case-based theory creation and variable directional hypotheses (SVH) and describes workable solutions to the mismatch. The solutions include case-based group-level analysis within symmetric testing (Bass et al. 1968), and resorting to simple algorithms to replace relying on symmetric tests (McClelland 1998) and using asymmetric Boolean-algebra-based indexes instead of symmetric tests (Ragin 2008). (Asymmetric tests are one-directional tests to indicate whether or not high scores in a complex state of *X* associate consistently with high scores in *Y*, an outcome condition. Chapter 2 explains asymmetric tests in some detail.) This chapter builds from a foundational premise, "'Scientists' tools are not neutral" (Gigerenzer 1991). Thus, if you test SVH via symmetric tests, you are not testing case-level causal conditions of complex wholes; you are testing a theory of net-effect directional hypotheses. SVH theory and tests include severe limitations beyond the mismatch between case-based theory and SVH. This chapter reviews these severe limitations (cf. Fiss 2007).

Following this introductory section, Sect. 2 focuses directly on overcoming the limitations of symmetric tests (regression analysis) and indirectly on the mismatch between case identification hypotheses (CIH) and SVH; Sect. 2 also describes one of the "illusions in regression analysis" (Armstrong 2012). Section 3 describes McClelland's algorithm procedure for overcoming the mismatch without completely moving beyond symmetric testing. Section 4 describes moving completely beyond symmetric tests and SVH to the use of asymmetric theory construction and tests of CIH. Section 4 describes a set of data to practice SVH versus CIH. Section 5 briefly describes three examples of matching asymmetric case-based theory construction with asymmetric CIH testing. Section 6 includes a visual and general discussion of case-based theory construction and CIH testing versus the conventional (and dominant) logic of boxes and arrows presentation of SVH representations of case-based theory. Section 7 concludes with a call for finance, management, and marketing scholars to recognize the current pervasive mismatch between case-based (typology) theories with the use of symmetric variable hypotheses testing—a mismatch that does not need to continue.

2 Symmetric Testing of Configural Outcomes to Overcome Regression Analysis Limitations

Most empirically based studies include symmetric (two-directional) variable hypotheses (SVH). The general form of these hypotheses includes the following expressions: (1) Increases in *X* (independent variable) associate with increases (decreases) in *Y* (dependent variable). (2) Increases in *W* (a second independent variable) associate with increases (decreases) in Y. (3) The impact of changes in *X* on *Y* is greater than the impact of changes in *W* on *Y*. (4) An interaction effect (*X* by *W*) occurs whereby the joint impact of increases in *W* and *X* is greater than the separate impacts of the levels of *W* and *X* on the level of *Y*. Researchers use symmetric tests such as analysis of variance and regression analysis to confirm or reject these expressions for variables in a given data set. Thus, the hypotheses are written in the following formats (with β representing a standardized partial regression coefficient of influence of change in *Y* due to a change in the independent variable):

1. $Y = \beta_1 X$, with $|\beta_1| > 0$, estimating the net effect of level of X on level of Y.
2. $Y = ß_2 W$, with $|\beta_2| > 0$, estimating the net effect of level of W on level of Y.
3. $Y = \beta_1 X + \beta_2 W$, where $|\beta 1| > |\beta 2|$, comparing the relative sizes of net effects.
4. $Y = \beta_1 X + \beta_2 W + \beta_3(X \bullet W)$, with all three $|\beta_i| \geq 0$, testing for an interaction effect beyond the net-effect influence.

In an article receiving a substantial number of citations (n = 200+) since its publication but widely ignored in practice, Bass, Tigert, and Lonsdale (1968) point to the severe limitations of using testing by symmetric metrics only. Their study, "Market Segmentation: Group Versus Individual Behavior," points out, "The evidence is overwhelming that R^2 (explained variance) is low when individual household purchase rates are related to socioeconomic variables. The intuitive conclusion, perhaps, suggested by the evidence is that market segmentation based on socioeconomic measurement is infeasible. This is the conclusion of Twedt (1964), Frank (1967) and others" (Bass et al. 1968, p. 264). Bass et al. (1968) go on to quote a few highly cited papers that include this inaccurate conclusion.

> The regression models applied in several studies of market segmentation have tended to focus on individual behavior, resulting in misleading conclusions. The propriety of the linearity assumption and noncontinuous observations on the dependent and independent variables brings into question the meaning of the regression results. This was essentially the point which Kuehn (1963) made in his debate with Evans (1959; Evans and Roberts 1963). Kuehn argued that simple cross-classification and analysis of a single variable with the probability of ownership was more revealing than discriminant analysis. It is not suggested that regression models are necessarily inappropriate for analysis of market segments. However, the results of these models should be interpreted in terms of group means (Bass et al. 1968, p. 265).

Bass et al. (1968) suggest that adopting a group segmenting procedure can show how socioeconomic variables do influence purchase behavior. Their procedure includes computing an estimated regression using the means for consumers in specific cells of a "segmentation structure" with each mean weighted by number of persons in the cell as the dependent variable. See Table 1.1 for an example of the means beer purchases for an income by education segmentation structure (i.e., cross-tabulation of respondents by 6 income levels by 5 education levels) that appears as Table 7 in their 1968 article. While not showing the actual empirical model, Bass et al. (1968, p. 269) report that for the estimated regression model, "… R is equal to .65 [the adjusted R^2 is close to 0.42], and the regression coefficients for both variables are substantially larger than their standard deviations. Thus when the noise is eliminated from the data, it is even more obvious that variables effectively discriminant between groups with different mean purchase rates."

Table 1.2 shows a reconfiguration of the data in Table 1.1 and the findings of estimating regression models using the means for both independent variables. For the two-term model, the adjusted R^2 equal to 0.38 is a bit lower than 0.42 that Bass et al. (1968) report most likely due to using unweighted means (30 rows of data) rather than weighted means (1400 rows of data). The adjusted R^2 equal to 0.38 is impressive and illustrates the benefit of using the mean for the dependent variable to represent all values in a given cell with a cross-tab of two or more independent variables.

Table 1.1 Cross-classification analysis of beer purchase of 1400 households and education by income

		Years of Education				
Annual family income		6	10	12	14	16
Under $3	Mean	10.01	6.53	6.18	12.27	15.21
	s.e.	1.64	2.38	1.74	7.78	13.17
	n	124	36	38	7	4
$3–4.999	Mean	27.74	20.27	11.70	17.40	1.79
	s.e.	3.93	4.03	2.98	9.51	1.65
	n	48	38	45	14	7
$5–7.999	Mean	25.23	26.03	22.63	24.27	16.80
	s.e.	2.62	2.45	1.85	3.53	3.85
	n	115	122	196	57	35
$8–9.999	Mean	27.72	24.21	32.14	21.78	23.23
	s.e.	6.47	4.51	3.07	3.92	3.78
	n	30	56	88	32	30
$10–14.999	Mean	34.24	24.05	21.54	30.63	24.18
	s.e.	6.47	4.51	3.07	3.92	3.78
	n	15	37	61	45	50
$15+	Mean	36.58	12.50	28.49	34.17	17.86
	s.e.	10.68	0.00	6.93	8.52	3.80
	n	7	1	15	10	37

Notes. Data are for 1964. *s.e.* standard error of the mean. *n* number in sample
Source: Data appears in Table 7 in Bass et al. (1968, p. 270)

However, the Bass et al. (1968) procedure has severe shortcomings. The procedure Bass et al. (1968) use eliminates more than just noise in each cell. The procedure eliminates the possible extraction of additional information from each cell in the cross-tab. In Table 1.1, most cells, including the two cells having the highest mean beer purchases, include non-beer purchase cases, that is, cases contrary to the main finding of high and low beer purchases occur in nearly every cell. For example, consider the cell having the highest mean beer purchases—bottom left corner in Table 1.1—the cell with highest income and lowest education level. The cell mean equals 36.58; the standard error (se) equals 10.68 for the 7 cases in this cell. Consequently, the standard deviation (s) equals 28.26 (se = s / $n^{.05}$; 10.58 = s / $7^{.05}$); solving for s equals 28.26. The large value for s relative to the mean indicates that more than one case with very low non-beer purchases occur in this cell. Rather than noise, a case identification hypotheses (CIH) perspective would seek to model these cases as well. Who are the contrarian consumers to the finding of very high beer purchases among the cases who are very high in income and very low in education? Answering this question is possible, but regression analysis is a poor tool for such digging.

Additionally, because mean values are used for all cases in specific cells, the Bass et al. (1968) procedure will not fare well in comparison to alternative procedures to accurately predict scores in additional samples. The prediction for specific scores will be far off the mark for contrarian cases using this mean centering procedure. Given that contrarian cases will occur for both directions in symmetric tests, the resulting adjusted R^2 will be substantially lower in tests on additional samples using the Bass et al. (1968) modified regression analysis method. Similarly almost all published studies using regression analysis that Bass et al. (1968) discuss use fit validity only and do not examine the predictive validity of their model for beer purchases or other product purchases. Studies consistently report that the fit is not a good way of assessing predictive ability (e.g., Gigerenzer and Brighton 2009; Pant and Starbuck 1990). "The obvious solution is to avoid the use of t, p, F, R^2 and the like when using regressions" (Armstrong 2012, p. 29).

The use of regression analysis, a symmetric variable directional hypotheses (SVH) testing tool, in market segmentation studies includes a second severe limitation. As a symmetric tool, regression analysis tests models that predict both low and high scores for a dependent variable. Market segmentation theory focuses on case identification theory and hypotheses (CIH)—using unidirectional theoretical statements where unidirectional tools work best. The title of Twedt's (1964) iconic study illustrates the point, "How important to marketing strategy is the 'heavy user'?" The foundational issue in market segmentation research focuses on describing and accurately predicting the heavy user—a unidirectional issue. The use of regression analysis-based theory and methods for a unidirectional issue is a mismatch of theory and data analysis.

Even though the authors suggest a modified regression analysis procedure, recognition is apparent in Bass et al. (1968) that regression analysis is not as useful as alternative tools for identifying cases of particular interest; they provide findings from a three-variable cross-tab for identifying heavy beer purchasers. Table 1.4 in Bass et al. (1968) identifies consumer cases of heavy beer purchases to be persons between 25 and 50, not college graduates, and watching more than 3.5 hours of TV per day. This description is an asymmetric unidirectional description—a case-level recipe that includes high scores in all three ingredients. This implication is that cases (i.e., consumers) having high scores in all three ingredients are heavy purchasers of beer, consistently. "Consistently" expresses the perspective that nearly all cases having high scores in all three ingredients are heavy purchases of beer. In real life, identifying cases having the specific levels for three ingredients is unlikely to be sufficient for high accuracy in predicting heavy beer purchases for additional samples. However, a researcher is likely to identify high beer purchases consistently by adding two additional ingredients into the recipes: males and sports watching. Thus, males between 25 and 50, not college grads, and watching sports on TV 3.5 hours per day are likely to be high beer purchasers consistently. This statement does not suggest necessity but only sufficiency. Certainly, some persons with a low score in on or more of these ingredients are heavy beer purchasers. The statement is a unidirectional statement of sufficiency only. We return to the statement and beer in the following sections.

Usually researchers seek to create models that consistently identify the heavy users, top-performing firms, or the highly competent executives; more generally, they seek to identify the top quintile (20%) or top decile (10%) cases of excellent firms or achieving executives. Research on strategic firm types is one example; Fiss (2011) constructs unidirectional models of antecedent recipes to identify firms con-

Table 1.2 Income, education, gender, and beer purchases: hypothetical data for first 22 of 60 cases

Case	Income	Education	Gender	Beer	inc_c	edu_c	beer_c	not_ edu_c	gen_inc_ notedu_c
Linda	1500	6	0	10	0.01	0.03	0.07	0.97	0
Jeff	1500	10	1	7	0.01	0.23	0.02	0.77	0.01
Mary	1500	12	0	6	0.01	0.5	0.02	0.5	0
Tom	1500	14	1	12	0.01	0.95	0.13	0.05	0.01
Phil	1500	16	1	15	0.01	1	0.32	0	0
Carol	4000	6	0	27	0.05	0.03	0.98	0.97	0
Bruce	4000	10	1	20	0.05	0.23	0.75	0.77	0.05
Arnold	4000	12	1	12	0.05	0.5	0.13	0.5	0.05
Donna	4000	14	0	17	0.05	0.95	0.5	0.05	0
Deb	4000	16	0	2	0.05	1	0	0	0
Lance	6500	6	1	25	0.27	0.03	0.95	0.97	0.27
Sandra	6500	10	0	26	0.27	0.23	0.97	0.77	0
Kathy	6500	12	0	23	0.27	0.5	0.9	0.5	0
Bill	6500	14	1	24	0.27	0.95	0.93	0.05	0.05
Ralph	6500	16	1	17	0.27	1	0.5	0	0
Peter	9000	6	1	28	0.69	0.03	0.98	0.97	0.69
Sue	9000	10	0	24	0.69	0.23	0.93	0.77	0
Matt	9000	12	1	32	0.69	0.5	1	0.5	0.5
Jane	9000	14	0	22	0.69	0.95	0.87	0.05	0
Martha	9000	16	0	23	0.69	1	0.9	0	0
Carl	12,500	6	1	25	0.95	0.03	0.95	0.97	0.95

sistently with high prospector scores—a firm strategic type. Fiss (2011) proposes other unidirectional recipe models to identify firms high in analyzer firm scores and additional models for firms high in defender firm scores. However, unlike Fiss (2011) who matches method and theory, most researchers make use of regression analysis tools and mismatch theoretical objectives with data analytical tools.

An interaction term was also constructed to test a three-term regression model for the data in Table 1.2. Income values were multiplied by education to construct this third variable. Using the enter command in MRA in the software statistical package (SPSS) resulted in a slightly lower adjusted R^2 equal to 0.36 with none of the *b* coefficients statistically significant. None of the partial nonstandardized regression weights are significant in this model because of high multicollinearity—both income and education correlate significantly with the interaction scores for these two variables. Quite frequently, researchers will report empirical regression models showing nonsignificant *b* coefficients with high adjusted R^2s and conclude falsely that the variables in these models have small net-effect sizes. The use of the stepwise command in MRA using SPSS resulted in only income entering the model and an adjusted R^2 equal to 0.31. Armstrong (2012, p. 690) discusses these illusions and recommends testing simple a priori reasoned regression models only and "do not to estimate relationships for more than three variables in a regression." The findings of Goldstein and Gigerenzer (2009) are consistent with this rule of thumb.

Armstrong (2012) also describes the illusion of control which appears frequently in regression models with five to twenty or more terms. Users of regressions assume that by putting variables into the equation, they are somehow controlling for these variables. However, this control only occurs for experimental data. Adding variables in a regression model does not mean controlling for them in nonexperimental data because many variables typically co-vary with other predictor variables. The problem becomes worse as more variables are added to the regression. Large sample sizes cannot resolve this problem, so statistics on the numbers of degrees of freedom are misleading (Armstrong 2012).

3 Use of Algorithms of Configural Antecedents to Overcome Regression Analysis Limitations

McClelland (1998) describes the use of algorithms of configurations of antecedent conditions to overcome the shortcomings of regression analysis. McClelland (1998) was trying to forecast outstanding (O) versus typical (T) executives using eleven "behavioral event interview" (BEI) competencies. "Generally, the O group is the top 5% to 105 of the executives [in a firm or other organization], and the T group includes the next 11–25% of the executive" (McClelland 1998, p. 332). The eleven BEI competencies were specifically designed interviews where members of the O and T groups describe in their own words, what they said, thought, felt, and did in six episodes—three positive and three negative—at work. The competencies were refined to improve the degree to which they distinguished between O and T executives. "If a competency is found to differentiate these two groups across samples of executives, it becomes part of a standardized dictionary of competencies" (McClelland 1998, p. 332). Here are the top four competencies in distinguishing between O and T executives: achievement orientation, analytical thinking, developing others, and impact/influence (from Table 1.1 in McClelland 1998, p. 332).

McClelland ran a series of additive regression models using the standardized competencies to predict different dependent variables (e.g., annual bonus awards to executives and O versus T memberships were dependent variables). McClelland reports that the additive regression models did not yield stable results.

> In one regression analysis, the previously validated competencies yielded a multiple R of .52 ($p < .10$) with bonus criterion [dependent variable]; however, in a second sample of 42 similar executives, the same regression formula failed to predict the same criterion ($R = .24$, n.s.). In this latter sample, a regression model based on different competencies did predict the criterion ($R = .58$, $p < .01$), but this model yielded an R of only .08 when applied back to the original sample (McClelland 1998, p. 334).

For each sample of executives, McClelland first tested a model for its fit validity and then tested the accuracy of the model on an additional sample of executives for predictive validity. He also tested a model developed using data in a different sample of executives for fit validity and then tested the model for predictive validity on the first sample. This dual testing for fit and predictive validation on two samples is "cross-validation" of a model.

Unfortunately, predictive validation of a model by testing using additional samples of cases is rarely done by researchers though predictive validation when done shows that regression models do poorly in comparison to simple algorithm models (Gigerenzer and Brighton 2009). Roberts and Pashler (2000) estimate that, in psychology alone, the number of articles relying on a good fit as the only indication of a good model runs into the thousands—the same observation applies for almost all articles in finance, management, and marketing journals using symmetric tests (correlation, MRA, and SEM). Good fit is easy to achieve as Armstrong (1970, 2012) demonstrates. Armstrong (2012) refers to running a regression model and reporting the resulting model (i.e., fit validity) to be the fit that implies accuracy illusion. Armstrong (1970) shows that achieving high fit validity is possible using a table of random numbers as a data set.

> In one of my Tom Swift studies, Tom used standard procedures, starting with 31 observations and 30 potential variables. He used stepwise regression and only included variables where [Student] t was greater than 2.0. Along the way, he [Tom Swift; Armstrong's doppelgänger] dropped three outliers. The final regression had eight variables and an R^2 (adjusted for degrees of freedom) of 0.85. Not bad, considering that the data were from Rand's book of random numbers (Armstrong 1970, p. 691).

Czerlinski, Gigerenzer, and Goldstein (1999), Gigerenzer and Brighton (2009), and Zeller (2001) demonstrate that multiple regression excels in data fitting ("hindsight"), that is, fitting its parameters to data that are already known, but performs relatively poorly in prediction ("foresight," as in cross-validation). Typically, fit improves as complexity in models increase (such as adding more independent variables), while the ex-ante forecast (i.e., predictive validation with an additional sample) accuracy decreases—a conclusion that Zellner (2001) traces back to Sir Harold Jeffreys in the 1930s (Armstrong 2012). The pervasive use of regression analysis in the behavioral sciences and sub-disciplines of business—with reports of fit validity models only—is half of the grand incompetency in theory building and testing. The other half of the grand incompetency is the mismatch between case identification theory building and symmetric (two-directional) hypotheses testing. The two incompetencies are grand for the perspective of their pervasiveness in the articles appearing in scholarly journals.

McClelland observed that many relationships between success (a unidirectional concern) and the frequencies of different competencies were not linear and these relationships are not described well by correlation coefficients (a symmetric—two directional—test of association). Instead "tipping points" (Gladwell 1996) describe these relationships well, that is, the identification of an outcome of interest associates consistently only with certain configurations of levels of two or more antecedent conditions. McClelland observed that most (55%) outstanding (O) executives displayed 8+ different competencies in behavioral event interviews (BEIs). He also observed that few (20%) typical (T) executives displayed 8+ different competencies in the BEIs.

McClelland (1998, p. 334) recognized the necessity of conducting predictive validation using additional samples: "The critical question was whether the competency algorithm predicted actual performance in additional samples." Thus, using the 8+ competency algorithm, he was able to correctly identify most top performers in three separate samples of top performers: 11 of 17, 9 of 11, and 6 of 7 outstanding executives. In additional samples, McClelland reports that most typical executives exhibited fewer than 8 competencies in the BEIs in each of 3 samples: 23 of 25, 7 of 8, and 10 of 11 typical executives had scores below 8 combined competencies in the BEIs. Thus, the specific algorithm—"if the executive exhibits 8+ competencies, predict the executive's performance to be outstanding"—was found to have high accuracy (not perfect) in two-directional tests in predicting Os and Ts in additional samples beyond the sample used to create the algorithm. The algorithm replaces the use of regression analysis and avoids the use of estimating weights for the individual eleven competencies.

McClelland (1998) reports an additional examination of BEI data for different senior executives of a global firm resulting in his construction of the following somewhat more complex algorithm of BEI competencies. For identifying a senior executive as an O, the senior executive had to exhibit at least 1 of 3 individual-initiative competencies, at least 1 of 3 organizational-skill competencies, and a total of 6 out of 12 competencies that either most commonly differentiate significantly between O and T executives or are unique to the firm. This complex algorithm represents a unidirectional hypothesis of a case identification theory. An executive who achieves the three parts of this algorithm will perform at the O level. This statement indicates sufficiency but not necessity; some other executives not achieving the three-part algorithm may perform at the O level and some may perform at the T level; thus, the claim is not made that achieving all parts of the algorithm is necessary for future O performance. The algorithm states that achieving all three parts of the algorithm is sufficient for identifying high O executives consistently. Given that some executives are likely to have high future performances who do not achieve all three parts of the algorithm, more than this algorithm alone is necessary for identifying these other high-performing executives. An algorithm is but one screening mechanism for identifying O's; other screening mechanisms may work equally well or better.

The algorithm method that McClelland (1998) describes is a paradigm shift away from relying only on symmetric tests (e.g., regression analysis). While McClelland does not offer a general theory for this shift, his insights and data analytics are useful steps toward more formal theory construction using asymmetric reasoning. The next section describes these advances.

4 Asymmetric Theory and Data Analysis

In part, the basic tenets of asymmetric theory building rely on core propositions in complexity theory (Wu et al. 2014). The basic tenets of asymmetric theory include the following complexity theory precepts.

Large effect size for a simple antecedent condition is neither sufficient nor necessary A specific level of a simple antecedent condition is insufficient for accurately predicting the level of an outcome condition. Thus, watching sports competitions on TV frequently by itself is not an accurate predictor of heavy beer purchases even if the correlation between the two behaviors is positive and statistically significant (e.g., $r = .53$, $p < .001$). Simple antecedent conditions associate consistently only rarely with a given level of an outcome condition (e.g., a sufficient symmetric association indicates a very high correlation, e.g., $r \geq .80$, and looks like panel B in Fig. 1.1).

A few recipes of antecedent conditions are sufficient The performance of a specific level of an outcome condition (e.g., high firm success, heavy beer purchases, monthly holiday trip taking) depends on specific recipes of antecedent conditions. Such complex antecedence conditions are configurations of simple conditions. Bass et al. (1968, 267) present recipes of heavy buyers for nine product categories in their Table 1.4. For example, persons "between 25 and 50, not college grad, and watch TV more than 3.5 hours" are heavy beer buyers; persons with "incomes 10,000 or over [high incomes for the year of the data collection], with high score or less education" are heavy cream shampoo buyers.

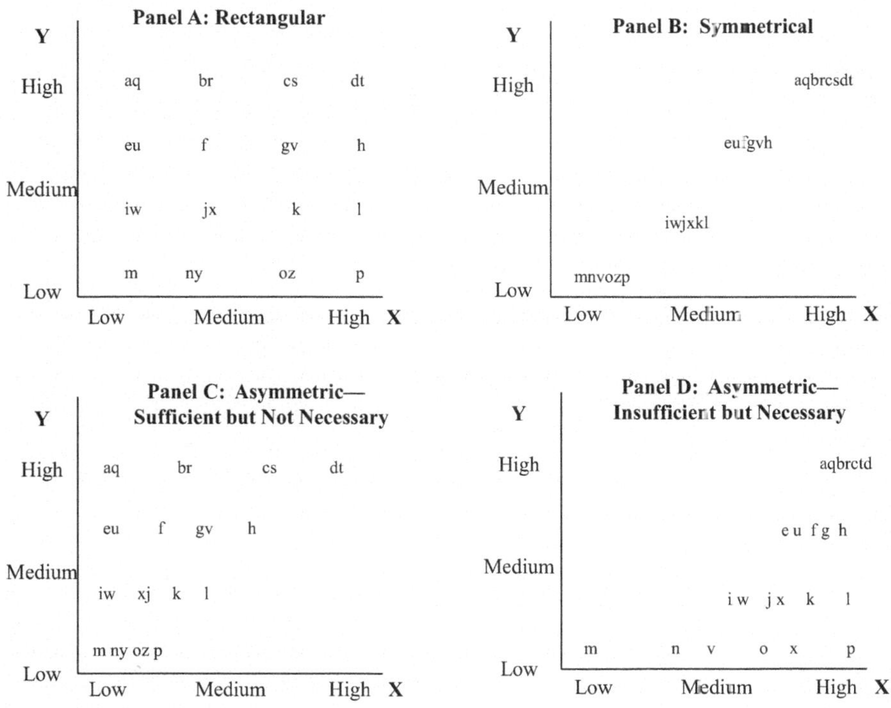

Fig. 1.1 Hypothetical relationships where *X* is a complex configural condition (e.g., C•T•H•Z) and *Y* is a service outcome condition (e.g., heavy beer purchases)

Some antecedent conditions are necessary but insufficient for identifying high *Y* Most simple conditions are neither necessary nor sufficient alone in predicting a level of an outcome condition. However, some conditions may be necessary but insufficient; for example, persons who are heavy beer consumption on a daily bases are all likely to be 25+ years of age. Panel D in Fig. 1.1 shows an example of a necessary but insufficient condition—some persons 25+ are heavy beer purchasers and some are not and all heavy beer purchases are 25+ years of age.

A simple antecedent condition can associate both with high and low levels of *Y* For many simple antecedent conditions, high levels of an antecedent condition associate with both a specific outcome condition and the negation of the same outcome condition—which depends on what additional ingredients occur in specific recipes of complex antecedent conditions associating with the specific outcome or the negation of this outcome. For example, frequent watching of sports competitions on TV may associate with heavy beer purchases as well as not buying beer at all.

The equifinality principle Different recipes of complex antecedent conditions are adequate for explaining a given level of an outcome conditions. This tenet is known as the equifinality principle—different routes (paths) are available that lead to the same outcome. For example, persons "between 25 and 50, not college grad, and watch TV more than 3.5 hours" are not all the people who are heavy beer purchasers.

The causal asymmetry principle Recipes associating with a low level of an outcome condition are not mirror opposites of recipes associating with a high level of the same outcome condition. This tenet is known as the causal asymmetry principle. For example, Bass et al. (1968, p. 267) describe light beer purchases to be "under 25 or over 50, college grad, nonprofessional, and watch TV less than 2 hours." Note that "nonprofessional" appears in this recipe, but occupational category does not appear in the recipe for the heavy beer purchaser.

Boolean algebra and fuzzy set qualitative comparative analysis (fsQCA) Unlike symmetric tests' reliance on matrix algebra, asymmetric theory construction and data analysis rely on Boolean algebra for specifying and testing recipes. Using Boolean algebra, the mathematical expression for a complex antecedent condition is equal to the lowest among the scores for the simple conditions in the complex statement. Thus, consider the following recipe and model for heavy beer purchases:

$$\text{income}\bullet \sim \text{education} \bullet \text{sports\_TV} \bullet \text{male} \;'' \; \text{heavy beer purchases} \qquad (1.1)$$

Model 1 claims that males with high incomes, low education (the sideways tilde, "~" represents negation), and who watch sports on TV frequently are heavy beer purchasers. The mid-level dot ("•") represents the logical "and" condition. Using fuzzy set membership scores for all the simple conditions in the model and the

Boolean algebra operation that the membership score for the recipe is equal to the lowest score among the simple conditions in the recipe, then the receipt membership score for a male with high income and very low education and who watched sports on TV every night would equal 1.0. For model 1 to be accurate, a person's membership score for beer purchases should also equal 1.0. For this model if this person is a female with the same high membership scores for the other simple antecedent conditions, the recipe score for this second person equals 0.00. Thus, for person 2:

$$\text{income} = 1.00; \sim \text{education} = 1.00; \text{sports\_TV} = 1.00; \text{male} = 0.00$$

complex antecedent recipe score equals 0.00. Both continuous and discrete values for conditions can be calibrated (i.e., transformed) into membership scores; membership scores run from 0.00 to 1.00.

The software program at fsQCA.com includes a subroutine for calibrating continuous values into membership scores for a logarithmic function (whereby values distant from the median are nearly equal to one another and values near the median are not equal to one another). The point here is that similar to Bass et al.'s (1968) attempt to eliminate noise, the fsQCA software for calibration removes the extreme influence of extreme scores. Thus, a $2.4 billion salary of one person (among 100 persons having a median salary of $80,000) would equal 1.0 for a calibrated membership score—the same membership score as the second highest salary (e.g., $700,000) among these 100 persons. Consequently, unlike using symmetric tests, no case "outliers" occur in asymmetric testing. Asymmetric testing via Boolean algebra versus symmetric tests via matrix algebra uses different concepts as well as mathematical operations. Table 1.3 presents a summary of these different concepts and operations.

Table 1.4 presents 22 cases of hypothetical data to illustrate raw values of variables and calibrated membership scores for cases for these values. Table 1.4 includes income, education, gender, and beer purchases for 22 cases. Linda is the first case in Table 1.4. Linda has the lowest income level ($1500), has completed sixth grade,

Table 1.3 Analogies and distinctions in using empirical positivism and fs/QCA

Empirical positivism	fsQCA
Independent variable	Antecedent condition
Dependent variable	Outcome condition
Interaction effect	Recipe
Objective 1: net effects of variables objective 1: multiple case ID recipe	
Objective 2: high adjusted R^2 highly consistent accuracy	
• = multiply	• = and
+ = add	+ = or
Value	Membership score
Z transformation score	Calibrated membership score
Two-directional tests	One-directional tests
Analysis: regression; ANOVA	Analysis: algorithms

Table 1.4 Computing consistency and coverage indexes

case	X (inc_~edu_gen_c)	Y (beer_c)	min (Xi,Yi)
linda	0	0.07	0
jeff	0.01	0.02	0.01
mary	0	0.02	0
tom	0.01	0.13	0.01
phil	0	0.32	0
carol	0	**0.98**	0
bruce	0.05	0.75	0.05
arnold	0.05	0.13	0.05
donna	0	0.5	0
deb	0	0	0
lance	0.27	**0.95**	0.27
sandra	0	**0.97**	0
kathy	0	0.9	0
bill	0.05	**0.93**	0.05
ralph	0	0.5	0
peter	0.69	**0.98**	0.69
sue	0	**0.93**	0
matt	0.5	1	0.5
jane	0	**0.87**	0
martha	0	0.9	0
carl	0.95	**0.95**	0.95
Sum	**2.58**	**12.8**	**2.58**

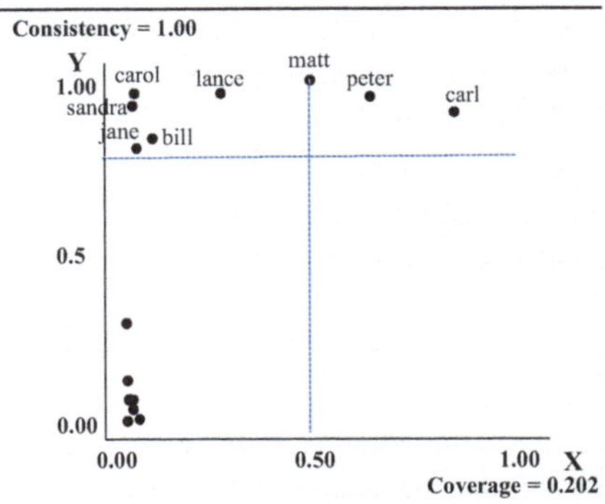

Consistency (Xi ≤ Yi)= Σ[min(Xi, Yi)]/Σ(Xi)
Consistency = 2.58/ 2.58 = 1.000

Coverage (Xi ≤ Yi)= Σ[min(Xi,Yi)]/Σ(Yi)
Coverage = 2.58/ 12.8 = .202

Note. **Heavy beer purchasing cases in bold and underlined.** The analysis indicates perfect (1.00) consistency and modest (.202) coverage. Data includes 8 heavy beer purchasers (≥.85); 3 of 8 have complex antecedent membership scores ≥ 0.5. Model does not explain Carol, Lance, Sandra, Jane, and Bill's heavy beer purchases; need to consider different antecedent models for them

and is female with a dummy score equal to 0.0 and beer purchases equal to 10. The calibrated membership scores for Linda also appear in the first row of data in Table 1.4. These membership scores include the calculation of ~education for Linda. Not education is the negation of education: .1–.03 = .97. The scores for the lowest education level could be calculated to be equal to 0.01 with no substantive change in the findings for this example. The calibrated scores reflect information for cases in a larger data file; the complete data file includes the data in Appendices 1 and 2 as well as Table 1.4. The estimated membership score for Linda for the model in the first column of Table 4 is equal to zero: income•~education•male (0.00) = .01•.97•.00 = .00 using Boolean algebra. The model tested here is that cases with high income, low education, and being male are heavy beer purchasers. Linda has low income, has low education, and is not male.

To examine a model's usefulness in identifying cases with high membership scores in an outcome condition, "consistency" and "coverage" indexes are estimated for different algorithm models via fsQCA. The consistency index indicates whether or not the model is dependable in accuracy in the membership scores for the simple or complex antecedent condition being equal or less than the membership score for the outcome condition across the cases in the study. The recommendation here is that the consistency index should be greater than .85 for model of antecedent conditions to be useful; the *XY* plot of the model should show the pattern apparent in panel C of Fig. 1.1. The coverage index estimates the relevancy of a model in estimating high membership scores in the outcome condition. The coverage index typically should range between .05 and higher. The equations for estimating consistency and coverage appear in Table 1.4 for the data for the 22 cases in Table 1.2.

The consistency and coverage findings in Table 1.4 indicate that high membership scores on the complex antecedents' model associate with heavy beer purchasers and that the model has perfect consistency. The findings indicate that high scores in the complex antecedent model are sufficient for concluding that the case represents a heavy beer purchaser, but additional models with high consistency are necessary for identifying additional heavy beer purchases. The model is sufficient but not necessary for identifying high scores on the *Y* (beer purchaser) condition. Do females who are heavy beer purchasers (drinkers) occur in real life? Yes, of course. However, the model tested here does not account for the female heavy beer purchasers. Additional models need to be proposed and tested that hopefully represent such female beer purchasers accurately.

Testing algorithms for predictive validity using additional samples Does the model just tested hold up well with case data in additional samples? Two additional samples of data appear in Appendices 1 and 2. For the data in Appendix 1, the model (income•~education•gender) has a consistency equal to .948 and coverage equal to .248; the model is quite useful for identifying cases of heavy beer purchasers. For the data in Appendix 2, the same model has a consistency equal to .767 and coverage equal to .651; the model is not useful for identifying cases of heavy beer purchasers. For the total sample combining Table 1.2 and the two appendices, the same model has a consistency equal to .881 and a coverage equal to .299; the model for the total cases ($n = 60$) is quite useful for identifying heavy beer purchasers. Panel D in Fig. 1.2 shows an asymmetric association between the complex antecedent model under discussion and heavy beer purchasers. Panels A, B, and C are for the simple antecedent conditions for income, education, and gender, respectively, and heavy beer purchasers. These three simple antecedent models are low in consistency.

Table 1.5 presents the findings from running a regression analysis—a symmetric test of how well the three independent variables predict low and high beer purchases. The symmetric variable hypotheses (SVH) test is useful with an adjusted R^2 equal to .267. However, unlike the algorithm case identification hypotheses (CIH), the SVH symmetric test does not identify heavy users and attempts to explain both directions in beer purchases—light/nonuse versus heavy use. Contrarian cases occur in everyday life; some females are heavy beer purchasers; some highly educated consumers are heavy beer purchasers; some low-income consumers are heavy beer purchasers—even when the main effects counter to these contrarian cases are significant statistically. The usual presence of contrary cases means that correlations between independent variables and a dependent variable will be modest even if the main effects are quite large. Constructing theory to test

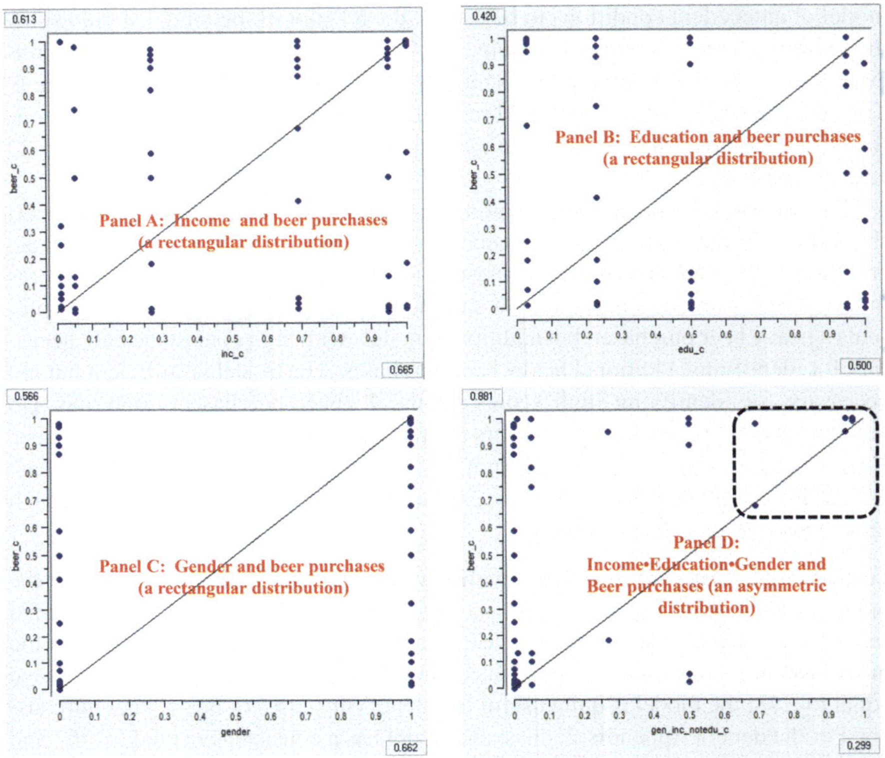

Fig. 1.2 Rectangular and asymmetric findings for three simple antecedents and gender•income•educataion ≤ beer purchases

for identifying cases with high scores on the outcome condition and using asymmetric tests to confirm or refute the theory is closer to reality and closer to what researchers typically seek to study.

5 Research Applications of Case-Based Theory and Case Identification Hypotheses Testing

Several examples of case-based theory matched with case identification hypotheses testing are available in the literatures in business (management) sub-disciplines. Fiss (2011) makes use of survey data from 205 high-technology manufacturing firms located in the United Kingdom to examine complex causal recipes that associate with high and very high firm performances. For very high performance, Fiss used the calibrated scores including 0.00 for average or below-average performance (ROA 7.8%, i.e., about the 50th percentile) and 1.00 for an ROA of 25% and above; and a crossover point for very high performance, he selected a calibrated score

Table 1.5 Summary of symmetric test of main effects

Regression statistics		Correlations	Income	Edu	Gende	
Multiple R	0.551	Income	1			
R square	0.304	edu	.000	1		
Adjusted R square	0.267	gender	0.052	−0.109	1	
Standard error	9.089	beer	0.299	−0.335	0.369	
Observations	60					
ANOVA						
	df	SS	MS	F	Significance F	
Regression	3	2020	673.4	8.15	0.0001	
Residual	56	4626	82.6			
Total	59	6646				
	Coefficients	Standard error	*t*-stat	*P*-value	Lower 95%	Upper 95%
Intercept	18.443765417	4.793747035	3.846188	0.000309	8.834624915	28.04068343
Income	0.00054334	0.000214689	2.530821	0.014216	0.000113266	0.000973414
Education	−0.917586714	0.343054346	−2.67476	0.009785	−1.604807149	−0.230366279
Gender	6.798785557	2.369183291	2.869675	0.005787	2.052741118	11.54483

equal to 16.3% ROA as the outcome condition membership scores. In his Table 1.5, Fiss (2011, p. 409) reports that three complex antecedent strategy recipes identify firms with very high performances. High structure complexity appears as a core ingredient in two of the three recipes; the negation of structure complexity appears as a periphery ingredient in the third recipe—such findings support the complexity theory proposition that high *X* relates to high *Y*, and the negation of *X* may relate also to high *Y* in the same data set for different cases via different causal recipes. Low-cost strategy appears in all three recipes for identifying very high-performing firms.

Ordanini et al. (2014, p. 134) demonstrate that recipes "are more important than the ingredients" in their "Qualitative Comparative Analysis (QCA) of Service Innovation Configurations." They hired a professional telemarketing company to recruit a random sample of 300 customers who had stayed at a luxury hotel for leisure during the prior 3 years. Each customer received an e-mailed description of a new hotel service and was scheduled for an interview 3–5 days later. Data on the study's possible antecedent conditions were collected during the interviews. They varied three levels of coproduction requirements in a between-subjects design. New service outcome intentions were the outcome condition in their study. The findings provide strong support for the three propositions that their study examines.

Proposition 1: The same attribute can either foster or inhibit new service adoption, depending on how it is configured with other attributes.

Proposition 2: Disparate configurations of service attributes are equifinal in leading to adoption.

Proposition 3: For adoption to occur, a configuration of attributes must be perceived as fitting the coproduction requirements.

While not referring to the core tenets in complexity theory, Ordanini et al.'s (2014) three propositions and findings provide direct support for the core tenets. The Ordanini et al. study is a major breakthrough study in solving the paradoxical findings in the literature on adoption of product-service innovations.

> Our QCA application also helps reconcile the inconclusive evidence concerning the role of individual drivers of service innovation success (Szymanski, Kroff, and Troy 2007). Our findings suggest that relative advantage, by itself, is necessary—but not sufficient—for adoption, while the other three drivers—novelty, meaningfulness, and complexity—when considered individually, are neither necessary nor sufficient. What matters for adoption to occur is whether the drivers and the coproduction setting are appropriately aligned. (Ordanini et al. 2014, p. 144)

Ordanini et al. (2014) compare their case-based theory and CIH testing with the conventional and dominant logic approaches of case-based theory followed by SVH testing, for example, cluster analysis. From the findings for these comparisons, they conclude that the findings from QCA were richer and more precise than those from cluster analysis. "We repeated the clustering using latent class analysis but the results were still inferior to those obtained from QCA" (Ordanini et al. 2014, p. 141). They also modeled the negation of adoption intentions, and their findings indicate four complex antecedent models showing high non-adoption of the new service—these findings support the causal asymmetry principle in complexity theory.

Prado and Woodside (2015) provide a third example of matching case-based theory with testing CIH. This study tests a theory of causally complex configurations of antecedent conditions influencing the adoption versus non-adoption of international supplier ethical certification standards. Using objective measures of antecedents and outcomes, a large-scale study of exporting firms in the cut-flower industry in two South American countries (Colombia and Ecuador) supports the theory. Prado and Woodside (2015) include complex antecedent models for the adoption and the negation of adoption of supplier industry ethical standards. They conclude, "The findings support the general conclusion that a simple antecedent condition has a positive or negative influence on adoption or non-adoption is overly simplistic. Impacts of simple antecedent conditions depend on the presence, absence, or negation of additional simple antecedent conditions within alternative complex antecedent statements" (Prado and Woodside 2015, preprint copy).

6 Visualizing Matching Generalized Cased-Based Theory and Case Identification Hypotheses Testing

Figure 1.3 illustrates the use of Venn diagrams and causal recipes reflecting some of the core tenets of complexity theory. Figure 1.3 illustrates the idea of moving away from net-effect interpretations and that different antecedent variables have different sizes of net effects. The typical real-life finding is that no single variable is sufficient or necessary for predicting cases with high outcome scores consistently in almost all studies. For the models in Fig. 1.3, religiosity is an exception to the just-made assertion. High national religiosity associates with nations high in altruistic behavior and low national religiosity associates with nations low in altruistic behavior—a symmetric relationship; the ability to predict nations with high and low altruistic behavior improves dramatically with the inclusion of additional ingredients in the respective recipes (Ferguson et al. 2015).

Notice in Fig. 1.3 that the visual includes separate models for high and low outcome scores. Such a visual separation of outcomes illustrates the asymmetric nature of reality and causal asymmetry in general—the models for the negation of altruistic behavior are unique models rather than simply the mirror opposite of models for high altruistic behavior. The thickness of the arrows in Fig. 1.3 is to indicate that some models usually have higher coverage—relevant for more cases—than other models. Such findings occur usually for the sets of useful models for both high and low outcome conditions.

Models 1 and 2 in Fig. 1.3 include several ingredients that are counter to one another. For example, the negation of ID appears in model1 and high ID appears in

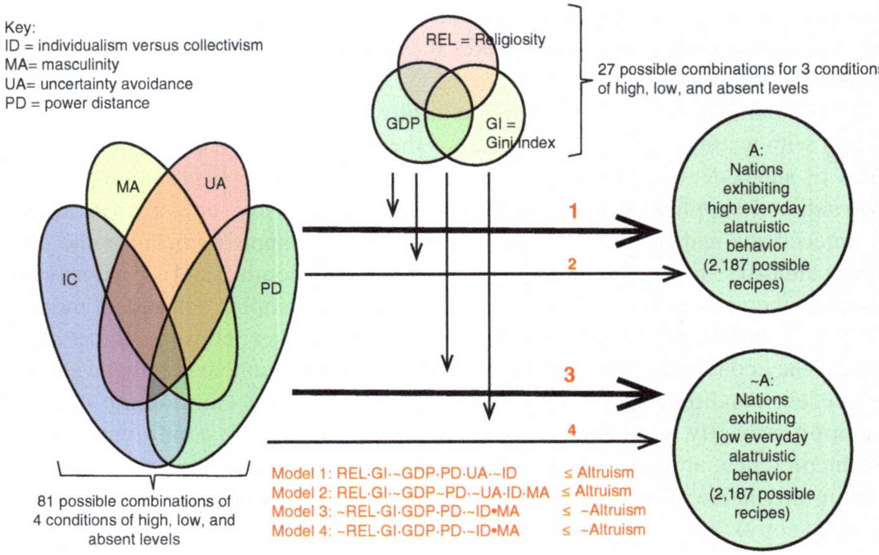

Fig. 1.3 Visual of case-based theory construction and case identification hypotheses testing. Notes. Visual here uses Hofstede's perspective that cultural values affect behavior but the visual is modifiable to represent other case-based theories with other moderating conditions affecting an outcome and negation outcome condition. This visual illustrates core tenets of complexity theory: two or more recipes associate with cases having high outcome scores; two or more recipes associate with cases having low outcome scores; some ingredients in recipes associate with both high and low outcome scores; moderates may be ingredients in some but not all recipes. Thick arrows indicate higher coverage while thin arrows indicate low coverage. Specific models for each arrow appear below the visual. (Source: Models are from Ferguson, Megehee, and Woodside (2015))

model 2—with both models being different recipes for nations high in altruistic behavior. MA is absent in model 1 and a positive ingredient in model 2; this finding illustrates the three possible levels for each simple condition as contributions in different recipes. This exposition ignores the possibility of a fourth level, that is, a curvilinear contribution of a simple antecedent condition into a recipe. This fourth possibility is beyond the scope of the present discussion.

The number of the possible configurations (2187) is quite large for recipes among seven antecedent conditions. The software program, fsQCA.com, can manage to process all possible combinations for seven antecedent conditions—much like SPSS software considering positive and negative additional terms in its stepwise regression routine. Care is necessary not to automatically test for all recipes; researchers should apply Armstrong's (2012) advice about not performing stepwise regression to automatically applying case-based data analysis for all possible configurations of antecedent conditions. A priori theory should drive the testing of CIH. However, this heuristic is easy to suggest but tough to adhere to in practice. Surprises occur in testing seemingly unlikely associations with high as well as outcome conditions, especially when testing for all combinations that offer both consistency (≥0.85) and a practical level of coverage (= 0.50) is available by a few clicks using the fsQCA software. This last observation reflects the perspective that theory and data analysis are two-way processes even if the focus is on asymmetric associations.

7 Conclusion

Given the literature review here, a shift away from the use of symmetric tests only in research in the sub-disciplines of business research is gaining some traction. However, most researchers in these sub-disciplines are still unaware of the mismatch in their theories of typologies and their writing of symmetric variable hypotheses. The use of symmetric tools such as ANOVA, correlation, and MRA/SEM alone fails to uncover the inherent complexity in relationships that occur in researchers' data sets. "Turning the tables: How case-oriented methods challenge variable-oriented methods" (Ragin 1997) is a useful place to start reading about the mostly unrecognized severe limitations of symmetric testing. Ragin (1997) is an open access paper only two clicks away.

The recognition of unsound practices in relying on symmetric tests began before Bass et al. (1968) as they illuminate, but most scholars continue to ignore the suggestions and worthwhile procedures and tools that Armstrong (2012), Gigerenzer and Brighton (2009), McClelland (1998), and Ragin (2008) describe. Given the establishment and success of the members of COMPASSS.ORG, the valuable case theory-analytical matching studies by Fiss (2011), Ordanini et al. (2014), and others (e.g., Wu et al. 2014), along with the publication of recent books in management (Fiss et al. 2013) and marketing (Meier and Donzé 2013) on case-based theory matched with case-based analytics, asymmetric theory building, and data analysis, are likely to be entering a tipping point at about the midpoint of the second decade in the twenty-first century. This shift in theory construction and data analytics is sufficiently radical as to generate substantial resistance by many reviewers at most journals. One step likely to help overcome this resistance is for scholars to include both conventional symmetric variable-based analysis and case-based analysis and findings in their journal submissions; the study by Ordanini et al. (2014) includes both alternative approaches. Acceptances of submissions of case-based analytical submissions are likely to continue to be low for a few additional years if authors do not follow Ordanini et al.'s (2014) dual-analytical practice.

Appendix 1: Data for Cases 23–45

Cases	Income	Education	Gender	Beer	inc_c	edu_c	beer_c	not_ edu_c	gen_inc_ notedu_c
Liz	12,500	10	0	26	0.95	0.23	0.97	0.77	0
Chuck	12,500	12	1	23	0.95	0.5	0.9	0.5	0.5
George	12,500	14	1	24	0.95	0.95	0.93	0.05	0.05
Sarah	12,500	16	0	17	0.95	1	0.5	0	0
Doug	18,000	6	1	37	1	0.03	1	0.97	0.97
Betty	18,000	10	0	13	1	0.23	0.18	0.77	0
Albert	18,000	12	1	28	1	0.5	0.98	0.5	0.5
Nigel	18,000	14	1	34	1	0.95	1	0.05	0.05
Shirley	18,000	16	0	18	1	1	0.59	0	0
Judy	1500	6	0	14	0.01	0.03	0.25	0.97	0
Luke	1500	10	1	39	0.01	0.23	1	0.77	0.01
Adel	1500	12	0	11	0.01	0.5	0.1	0.5	0
Mark	1500	14	1	3	0.01	0.95	0.01	0.05	0.01
Roger	1500	16	1	9	0.01	1	0.05	0	0
Ann	4000	6	0	5	0.05	0.03	0.01	0.97	0
Kane	4000	10	1	11	0.05	0.23	0.1	0.77	0.05
Able	4000	12	1	4	0.05	0.5	0.01	0.5	0.05
Julia	4000	14	0	0	0.05	0.95	0	0.05	0
Peggy	4000	16	0	0	0.05	1	0	0	0
Don	6500	6	1	13	0.27	0.03	0.18	0.97	0.27
Meg	6500	10	0	5	0.27	0.23	0.01	0.77	0
Virginia	6500	12	0	2	0.27	0.5	0	0.5	0
Tim	6500	14	1	21	0.27	0.95	0.82	0.05	0.05

Appendix 2: Data for Cases 46–60

Case	Income	Education	Gender	Beer	inc_c	edu_c	beer_c	not_ edu_c	gen_inc_ notedu_c
Hugh	6500	16	1	18	0.27	1	0.59	0	0
Arch	9000	6	1	19	0.69	0.03	0.68	0.97	0.69
Christine	9000	10	0	16	0.69	0.23	0.41	0.77	0
Dilbert	9000	12	1	9	0.69	0.5	0.05	0.5	0.5
Audrey	9000	14	0	3	0.69	0.95	0.01	0.05	0
Vivian	9000	16	0	8	0.69	1	0.03	0	0
Aaron	12,500	6	1	33	0.95	0.03	1	0.97	0.95
Olivia	12,500	10	0	5	0.95	0.23	0.01	0.77	0

(continued)

Case	Income	Education	Gender	Beer	inc_c	edu_c	beer_c	not_ edu_c	gen_inc_ notedu_c
Nick	12,500	12	1	7	0.95	0.5	0.02	0.5	0.5
Kent	12,500	14	1	12	0.95	0.95	0.13	0.05	0.05
Kim	12,500	16	0	0	0.95	1	0	0	0
Graham	18,000	6	1	31	1	0.03	0.99	0.97	0.97
Ruby	18,000	10	0	5	1	0.23	0.01	0.77	0
Brad	18,000	12	1	36	1	0.5	1	0.5	0.5
Clark	18,000	14	1	3	1	0.95	0.01	0.05	0.05
Amber	18,000	16	0	6	1	1	0.02	0	0

References

Armstrong, J. S. (1970). How to avoid exploratory research. *Journal of Advertising Research, 10*(4), 27–30.

Armstrong, J. S. (2012). Illusions in regression analysis. *International Journal of Forecasting, 28*, 689–694.

Bass, F. M., Tigert, D. J., & Lonsdale, R. T. (1968). Market segmentation: Group versus individual behavior. *Journal of Marketing Research, 5*, 264–270.

Cyert, R. M., & March, J. G. (1963). *A behavioral theory of the firm*. Englewood Cliffs: Prentice-Hall.

Czerlinski, J., Gigerenzer, G., & Goldstein, D. G. (1999). How good are simple heuristics? In G. Gigerenzer, P. M. Todd, & the ABC Research Group (Eds.), *Simple heuristics that make us smart* (pp. 97–118). New York: Oxford University Press.

Evans, F. B. (1959). Psychological and objective factors in the prediction of brand choice: Ford versus Chevrolet. *Journal of Business, 32*, 340–369.

Evans, F. B., & Roberts, H. V. (1963). Fords, Chevrolets and the problem of discrimination. *Journal of Business, 36*, 242–244.

Ferguson, G., Megehee, C. M., & Woodside, A.G. (2015). *Cultural recipe explanations of consumer altruistic and anti-altruistic behavior*. Working paper. Curtin School of Business, Bentley, Australia: Curtin University, 55 pages.

Fiss, P. C. (2007). A set-theoretic approach to organizational configurations. *The Academy of Management Review, 32*(2), 1180–1198.

Fiss, P. C. (2011). Building better casual theories: A fuzzy set approach to typologies in organizational research. *Academy of Management Journal, 54*(2), 393–420.

Fiss, P. C., Marx, A., & Cambré, B. (2013). Configurational theory and methods in organizational research: Introduction. In P. C. Fiss, B. Cambré, & A. Marx (Eds.), *Configurational theory and methods in organizational research* (Vol. 38). Bingley: Emerald.

Frank, R. E. (1967). Market segmentation research: Findings and implications. In F. B. Bass, C. W. King, & E. A. Pessemier (Eds.), *Applications of the sciences in marketing management*. New York: Wiley.

Gigerenzer, G. (1991). From tools to theory: A heuristic of discovery in cognitive psychology. *Psychological Review, 97*, 254–267.

Gigerenzer, G., & Brighton, H. (2009). Homo heuristics: Why biased minds make better inferences. *Topics in Cognitive Science, 1*, 107–143.

Goldstein, D. G., & Gigerenzer, G. (2009). Fast and frugal forecasting. *International Journal of Forecasting, 25*, 760–772.

McClelland, D. C. (1998). Identifying competencies with behavioral-event interviews. *Psychological Science, 9*, 331–339.

Meier, A., & Donzé, L. (Eds.), (2012). Fuzzy methods for customer relationship management and marketing: Applications and classifications, IGI Global, USA (2012), 1–15.

Meier, A., & Donzé, L. (Eds.). (2013). *Fuzzy methods for customer relationship management and marketing: Applications and classification*. Hershey: IGI Global.

Hofstede, G. (2001). *Culture's consequences, revised ed.* Beverly Hills: Sage.

Howard, J. A., & Sheth, J. N. (1969). *The theory of Buyer behavior*. New York: Wiley.

Hsu, S.-Y., Woodside, A. G., & Marshall, R. (2013). Critical tests of multiple theories of cultures' consequences: Comparing the usefulness of models by Hofstede, Inglehart and Baker, Schwartz, Steenkamp, as well as GDP and distance for explaining overseas tourism behavior. *Journal of Travel Research, 52*, 679–704.

Kuehn, A. A. (1963). Demonstration of the relationship between psychological factors and brand choice. *Journal of Business, 36*, 237–241.

Miles, R. E., & Snow, C. C. (1978). *Organizational strategy, structure, and process* . [A. D. Meyer, collaborator; H. J. Coleman Jr., contributor]. New York: McGraw Hill.

Ordanini, A., Parasuraman, A., & Rubera, G. (2014). When the recipe is more important than the ingredients: A qualitative comparative analysis (QCA) of service innovation configurations. *Journal of Service Research, 17*, 134–149.

Pant, P. N., & Starbuck, W. H. (1990). Innocents in the forest: Forecasting and research methods. *Journal of Management, 16*, 433–446.

Prado, A.M., & Woodside, A.G. (2015), Deepening understanding of certification adoption and non-adoption of international-supplier ethical standards. J. Bus. Ethics, 132, 105–125.

Ragin, C. C. (1997). Turning the tables: How case-oriented methods challenge variable-oriented methods. *Comparative Social Research, 16*, 27–42 . Available at http://isites.harvard.edu/fs/docs/icb.topic631340.files/Ragin%20%2D%2D%20Turning%20the%20Tables.pdf

Ragin, C. C. (2008). *Redesigning social inquiry: Fuzzy sets and beyond*. Chicago: Chicago University Press.

Roberts, S., & Pashler, H. (2000). How persuasive is a good fit? A comment on theory testing. *Psychological Review, 107*, 358–367.

Szymanski, D.M., Kroff, M.W., . Troy, L.C. (2007). Innovativeness and new product success: Insights from the cumulative evidence. Journal of the Academy of Marketing Science, 35, 35–52.

Twedt, D. K. (1964). How important to marketing strategy is the heavy user? *Journal of Marketing, 28*, 71–72.

Woodside, A. G. (2013). Moving beyond multiple regression analysis to algorithms: Calling for a paradigm shift from symmetric to asymmetric thinking in data analysis, and crafting theory. *Journal of Business Research, 66*, 463–472.

Woodside, A. G., Hsu, S.-Y., & Marshall, R. (2011). General theory of cultures' consequences on international tourism behavior. *Journal of Business Research, 64*, 785–799.

Wu, P. -L., Yeh, S. S., Huan, T. C., & Woodside, A. G. (2014). Applying complexity theory to deepen service dominant logic: Configural analysis of customer experience-and-outcome assessments of professional services for personal transformations. *Journal of Business Research, 67*, 1647–1670.

Zellner, A. (2001). Keep it sophisticatedly simple. In H. Keuzenkamp & M. McAleer (Eds.), *Simplicity, inference, and modelling: Keeping it sophisticatedly simple*. Cambridge: Cambridge University Press.

Chapter 2
Constructing Algorithms for Forecasting High (Low) Project Management Performance

Olajumoke A. Awe, Arch G. Woodside, Sridhar Nerur, and Edmund Prater

Abstract Applying complexity theory tenets, the study here provides a unique asymmetric modeling perspective for examining causal conditions indicating high (low) project management performance (PMP). Complexity theory tenets include (tenet 1) recognizing that the causal conditions resulting in high PMP frequently have different components (i.e., ingredients) than the causal conditions resulting in low PMP—adopting this perspective supports the usefulness of asymmetric rather than the currently pervasive symmetric approach to theory construction and empirical modeling. A second complexity theory tenet is that the same causal condition can foster, be irrelevant, or inhibit high PMP, depending on how it is configured with other causal conditions—thus, high knowledge management effectiveness (KME) by itself is neither a sufficient nor a necessary causal condition for indicating all cases of high PMP. A third tenet is that the disparate configurations of causal conditions are equifinal in leading to adoption. The study here constructs a general model and specific configurational propositions that include social capital, project management types, processes, and complexity as causal conditions indicating case outcomes of high versus low PMP. The study includes examining the model and propositions empirically using survey data on the causal conditions for completed projects ($n = 302$, US sample of product and service industrial firms). The findings support the perspective that high (as well as low) PMP depends on combination

The authors are grateful for the insightful comments that colleagues, Rouxelle De Villiers, Auckland University of Technology, and Kurt Hozak, Coastal Carolina University, provided on an earlier draft of this paper.

O. A. Awe (✉)
Coastal Carolina University, Wall College of Business of Administration, Conway, SC, USA
e-mail: oawe@coastal.edu

A. G. Woodside
Yonsei University, Yonsei Frontier Lab, Seoul, Republic of Korea
e-mail: arch.woodside@bc.edu

S. Nerur · E. Prater
Department of Information Systems and Operations Management, University of Texas, Arlington, TX, USA
e-mail: snerur@uta.edu; eprater@uta.edu

A. G. Woodside (ed.), *Accurate Case Outcome Modeling*,
https://doi.org/10.1007/978-3-030-26818-3_2

effects—not the additive or net effects of causal conditions. For project managers, adopting a configurational approach to the study of project outcomes can reveal which combinations of causal conditions consistently lead to high PMP as well as which combinations indicate low PMP and the conditions when high KME associates with low PMP.

Keywords Algorithms · Configurations · KME · Performance · Project management · Social capital

1 Introduction

Is "knowledge management effectiveness" (KME) a necessary and/or sufficient condition for projects to achieve high performance? Is achieving high social capital a necessary and/or sufficient condition for projects to achieve high performance? The study here contributes to the project management performance (PMP) literature by constructing case-based forecasting models and empirical testing answers to these and related questions. In answering the first two research questions, rather than applying the currently pervasive symmetric, variable-directional perspective to theory construction and testing via null hypothesis significance tests (NHST), the present study applies asymmetric, case-based outcome, configurational theory, modeling, and empirical tests via somewhat precise outcome testing (SPOT) (Fiss 2007; Woodside et al. 2018). The findings below support the theoretical perspective that projects achieving the complex antecedent condition including both high KME and high social capital are sufficient for accurately predicting high PMP consistently (i.e., KME•SocialCapital→PMP, where the mid-level dot "•" represents the logical Boolean algebra "AND" operation and the directional arrow, "→," indicates prediction). However, this configuration is an asymmetric observation and prediction, that is, the KME•Social_Capital condition is not necessary for a project to result in a high PMP outcome; additional complex configurations of antecedent conditions indicate high PMP and additional asymmetric models indicate low PMP. Though comparatively rare to the construction and testing of symmetric variable-directional hypotheses, prior asymmetric configurational theory construction and empirical research using Boolean algebra appears in the project management literature (Chereshnyuk et al. 2017; Young and Poon 2013); for example, Young and Poon (2013, p. 943) provide asymmetric configurational theory and "findings from a fuzzy set analysis" indicating "top management support—almost always is necessary and sometimes sufficient for success." The present study contributes to the nascent project management literature stream of asymmetric modeling and research.

Knowledge management effectiveness (KME) is how well an organization creates, stores, transfers, and reuses its knowledge (Song et al. 2007, 2008a, b). Knowledge management involves an effort to gain useful knowledge within the forganization by encouraging communication and the free flow of ideas between employees, work units, and business units. KME in the organization involves the integration of knowledge from different sources (Ramesh and Tiwana 1999), and it impacts process innovation and improvement, executive decision-making, and organization adaptation (Earl 2001). In the business environment, the difference between a successful firm and an unsuccessful organization can be the way the firm attempts to achieve team manage knowledge. As project team members interact within and across teams, they share and transform information which leads to the creation, storage, and transfer of knowledge. The present study includes the finding that projects high in KME are sufficient marginally for indicating successful PMP consistently. However, KME alone is one ingredient among a few in indicating high PMP. KME alone is not necessary for the occurrence of high PMP.

Regarding social capital in sociology and project management, Coleman (1988) suggests that a relationship network with strong connections resembles social capital. According to social capital theory relating to PMP, interpersonal skills and relationships have a direct influence and affect the contributions within a research team that influence performance (Stewart and Barrick 2000). "Social capital" is a configurational concept in project management that refers to innovation, transformative capacity, and knowledge integration in a cooperative team (Obstfeld 2005; Tiwana 2008) and channels resources toward the project team leader's network connections (Kao and Shen 2009). Research findings in the PMP literature support the view that the depth of a social network reflects the importance of a person acting as the information or knowledge source within his or her associated organization or society (Björk and Magnusson 2009; Tsai 2001). The present study includes in-depth surveying of persons acting as prime information sources in team-supported organizational projects in 302 firms (US sample).

Nearly all studies contributing to project management theory and research apply symmetric approaches in the study of variable relationships and test whether or not these relationships permit acceptance of alternative hypotheses using null hypothesis significance tests (NHSTs). Rather than testing whether or not variable relationships are statistically significantly different from zero and the relative effect sizes of these relationships, the study here builds and tests asymmetric algorithms (i.e., configurations or screens) of complex antecedent conditions that indicate high project management performances (PMPs)—the study includes building and testing separate screens for low PMPs. The research perspective here builds from a foundation of complexity theory pillars (i.e., tenets). These tenets provide propositions that are testable for consistency of their predictive accuracies with the objective of achieving "somewhat precise outcome testing (SPOT)"—that is, the predicted outcome occurs for nearly all the projects matching the conditions in a configuration.

"Scientists' tools are not neutral" (Gigerenzer 1991, p. 264) applies here. One consequence of Gigerenzer's (1991) dictum is that the construction of variable-directional hypothesis of main effects usually follows with the use of NHST and comparisons of relative sizes of impacts of independent variables. For example, much of the literature on "critical success factors" (CSF) in the project management literature focuses on reporting on the relative sizes (e.g., beta's) of impacts of alternative CSFs (e.g., Banihashemi et al. 2017; Kumar and Sharma 2017; Marzagão and Carvalho 2016; Pal et al. 2017). The relevant literature for constructing and testing for future theory in project management research includes recognizing the criticisms and calls for ending the practice of constructing and testing variable-directional hypothesis and using NHST (Armstrong 2012; Fiss 2007, 2011; Hubbard 2015; Wasserstein and Lazar 2017; Woodside 2013; Ziliak and McCloskey 2008). Unfortunately, while concluding correctly that, "By itself, a *p*-value does not provide a good measure of evidence regarding a model or hypothesis (Wasserstein and Lazar 2016, p. 132), the 2016 "American Statistical Association Statement on p-Values: Context, Process, and Purpose" provides scant information of what should researchers should do once they recognize that using NHST and reporting p-values constitute bad science practices. Fiss (2007, p. 207) provides a clear way forward,

> …many of the problems of empirical research on organizational configurations derive from a mismatch between methods and theory. Configurational theory suggests a clean break with the predominant linear paradigm. Rather than implying singular causation and linear relationships, a configurational approach assumes complex causality and nonlinear relationships where variables found to be causally related in one configuration may be unrelated or even inversely related in another. As a result, relationships between variables need not be symmetric and tend to involve synergistic effects that go beyond traditional bivariate interaction effects.

The configurational theory differs fundamentally in theory construction and empirical testing from the symmetric analysis.

Consequently, rather than focusing on the net effects of variable relationships, applying configurational theory in research on PMP rests on core tenets of complexity theory. Section 2 presents the core tenets of complexity theory. Section 3 presents the configurational theory of the antecedent conditions indicating high versus low PMP. Section 4 includes the method for scale development, data collection, and data analysis. Section 5 presents the findings. Section 6 is the discussion section and describes the study's contributions to project management theory and practice and limitations. Section 7 provides a general conclusion.

2 Core Tenets of Complexity Theory Applicable for Examining PMP Antecedent Conditions

An asymmetric view of reality is a core tenet of complexity theory, that is, the causal configurational conditions indicating high scores for an outcome condition include different ingredients than causal configurational conditions indicating low for an outcome condition. Consequently, rather than using symmetric analytic tools such as multiple regression analysis that predict both low and high scores for a dependent variable, asymmetric models predict only the high score for an outcome condition (in the language of asymmetric modeling, "outcome condition" appears in place of "dependent variable"). Figure 2.1 depicts an XY plot indicating a symmetric variable relationship (panel B).

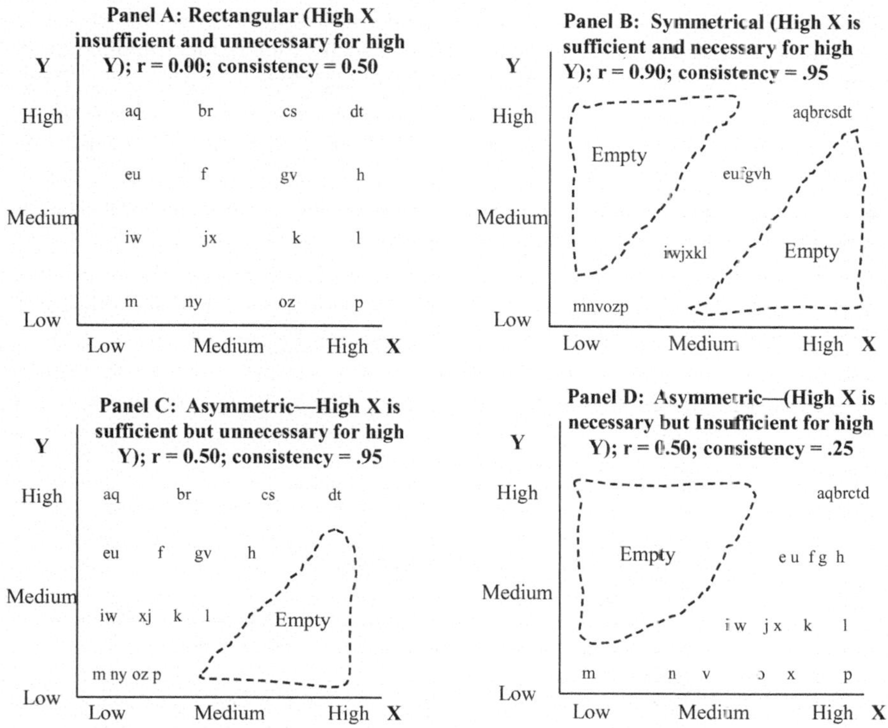

Fig. 2.1 Hypothetical relationships where X is a complex configural condition (e.g., social capital AND KME as X indicates high project management performance as Y)
Dictionary: KME = knowledge management effectiveness
Notes
Panel A: cases high in X are both high and low in Y. Findings are not symmetric nor asymmetric. For panel A, *r* correlation equals 0.00 and consistency (a measure for asymmetry) equals 0.50
Panel B illustrates symmetric theory and findings. Regression analysis and structural equation modeling (SEM) illustrate symmetric model construction and testing. For panel B, *r* = 0.98; consistency = 0.95
Panel C indicates an asymmetric relationship rather The present study specifies and tests for models that illustrate the pattern of cases appearing in panel C. Asymmetric modeling does not attempt to predict both low and high scores in an outcome condition (Y) but only predict high scores in the outcome condition (Y). For panel C, *r* = .50, consistency = 0.95. For consistency, high scores on X indicate high scores on Y
Panel D: some antecedent conditions are necessary but insufficient for indicating cases with high scores in Y (e.g., being "male" is a necessary but insufficient condition to indicate players (high Y) playing football in the US National Football League. For panel D, *r* = 0.50, consistency = 0.15. For panel D, high scores on X are necessary but insufficient for indicating high scores on Y

Figure 2.1 also depicts an XY plot indicating an asymmetric association for X and Y scores. Low scores for Y can be modeled via asymmetric analysis by computing the negation of Y. An asymmetric prediction is one directional: high scores in X indicate high scores in Y. An accurate asymmetric model indicates sufficiency that high X indicates high Y but not a necessity—low scores in X may associate with both low and high scores in Y and such a finding takes nothing away from the accuracy of the model's prediction about high X and high Y scores. Panel A in Fig. 2.1 shows a "rectangular" relationship between X and Y: cases low (high) for X are low (high) in Y. Panel D in Fig. 2.1 shows a necessary but insufficient association between X and Y: cases high in X are both low and high in Y, but no cases low in X have high Y scores. Some necessary but insufficient relationships (panel D) can be important while others are trivial. Building financial stock purchase and human resource hiring/promoting screens are examples of constructing asymmetric models of combinations of conditions sufficient for identifying cases whereby high X scores indicate high Y scores consistently (panel C in Fig. 2.1).

Table 2.1 includes five complexity theory tenets. The first tenet is the insufficiency tenet that represents the XY association whereby high X is necessary but

Table 2.1 Core tenets of complexity theory

Tenet	Concept	Description	Boolean Expression
T1	Insufficiency	High X may be necessary, but this condition is insufficient for identifying cases high in Y	X //→ Y (panel D in Fig. 2.1)
T2	Equifinality	A few, not one, distinctly unique complex configurations of antecedent conditions indicate the same outcome	(X•R) ≤ Y + (~X•T) ≤ Y (panel C in Fig. 2.1)
T3	Contrarian	Both high X and low X associate with high Y Both high X and low X associate with low Y	(X • R) ≤ Y + (~X • T) ≤ Y (X • W) ≤ ~Y + (~X • F) ≤ ~Y
T4	Causal asymmetry	Complex antecedent conditions for low Y are not the mirror opposite of complex antecedent conditions for high Y	(X • R) ≤ Y ≠ (~X • ~R) ≤ ~Y
T5	Emergence	System effects occurring in creating configurations of simple conditions are greater than the sum of the simple conditions (where SE = self-esteem, GSE = generalized self-efficacy, LC = locus of control, ES = emotional stability, and CSE = core self-evaluations)	(SE•GSE•LC•ES > CSE_{total})

Key: Boolean algebra operational meanings: mid-level dot, "•", indicates the logical "and"; sideways tilde, "~", indicates negation; the plus sign "+" indicates "or"; the less than or equal sign, "≤" and the directional arrow ("→") indicate scores for the model input statement are all or nearly all lower than scores for the outcome, Y or (Y • Z); the not equal sign, "//→," indicates that the input model (simple or complex) does not indicate an asymmetric pattern that screens for Y or ~Y where "Y" refers to cases with high Y scores and "~Y" refers to cases with low Y scores, the negation of a Y score; "X" refers to high X scores and "~X" refers to low X scores. X, R, F, and W refer to simple antecedent conditions; Y and Z refer to simple outcome conditions; "≠" refers to causal asymmetry

Notes: A useful heuristic is to discretize scores when calibrated values of a variable into fuzzy-set scores so that all cases in the lowest quintile have fuzzy scores ≤0.10 and cases in the highest quintile have fuzzy scores ≥0.90. Configural analysis and setting consistency requirements are "fuzzy" in deciding what constitutes low (e.g. ~Y) and high (Y) scores and in deciding on the limit necessary for models of complex antecedent conditions to surpass to indicate high accuracy in predicting Y or ~Y insufficient for identifying cases having high Y scores (panel D in Fig. 2.1). The second tenet in Table 2.1 states that an asymmetric model may be sufficient for identifying cases having high X scores that have high Y scores consistently—nearly always. Panel C in Fig. 2.1 represents this second tenet. A complex X is a configurational statement, for example, cases high in KME as well as being high in social capital—a specific project (case) must have high scores in both KME and social capital for a high X score. Original scores are calibrated using a 0.00 to 1.00 scale and the Boolean "AND" operation of using the lowest score for the simple conditions to represent the set score. Thus, if a project has a KME calibrated score equal to 0.35 and a social capital calibrated score equal to 0.95, the configurational score equals 0.35—the amount of scoring that the two conditions share. Consequently, for symmetric scoring combining two simple conditions into a complex score requires both scores to have high calibrated scores—for example, a second project having a KME score equal to 0.80 and a social capital score equal to 0.81, its complex score would equal 0.80.

The reality of contrarian cases to the variable-directional hypothesis is one way of expressing the third tenet of complexity theory. Even when an XY relationship in a study indicates a statistically significant finding rejecting the null hypothesis, contrarian cases usually occur—cases showing the opposite relationship in an XY plot or by discretizing both X and Y data and cross-tabbing using quintiles. Using quintiles, the findings in Table 2.2 from the present study for KME and PMP illustrate this tenet and finding. Cases exist in all four corners of the 5 by 5 cross-tabulation

Table 2.2 Cross-tabulation of case outcomes via discretizing using quintiles: knowledge management effectiveness (KME) and project management performance (PMP)

		Project management performance					
		Very low 1.00	2.00	3.00	4.00	Very high 5.00	Total
Knowledge management effectiveness	Very low 1.00	32	10	14	8	2	56
	2.00	17	24	11	11	6	59
	3.00	5	14	31	15	7	72
	4.00	8	8	6	35	8	65
	Very high 5.00	0	3	0	11	16	30
Total		62	59	62	80	39	302

Cases where:
~KME•~PMP: 27%
~KME• PMP: 9%
KME•~PMP: 5%
KME• PMP: 23%

Key: The mid-level dot, "•", indicates the Boolean logical "AND" operation; the sideways tilde, "~," indicates the negation of a calibrated score. The numbers in Table 2.2 indicate the number of cases occurring in each of the 25 cells in this 5-by-5 crosstab

Notes:

Phi = 0.714, $p < .000$ for this cross-tabulation of quintiles

$r = 0.54$, $p < .000$ using original continuous values of the data

Linear regression analysis: PMP = 2198.85 + (29.27•KME), for b coefficient, $t = 11.17$, $p < .000$

Even though the symmetric analysis indicates a highly statistical significant relationship (e.g., b coefficient = 29.27 supports the alternative hypothesis that $b \neq 0.00$), discretizing the data into quintiles and cross-tabulating KME by PMP indicate 14% of the cases are contrary to a positive main effect for KME and PMP

Asymmetric analysis includes modeling cases in all four corners: cases where high KME indicates high PMP; cases where high KME indicate low PMP; cases where low KME indicate high PMP; and cases where low KME indicate low PMP of KM and PMP in Table 2.2 even though the phi coefficient indicates a statistically significant positive relationship between X and Y (phi = 0.714, $p < 0.000$). Positive PMP outcomes occur for cases having low KME conditions for 9% of the 302 cases in the study (i.e., the cases in the upper-right cells in Table 2.2). Negative PMP outcomes occur for cases having high KME conditions for 5% of the 302 cases (i.e., the cases in the lower-left cells in Table 2.2). Adopting an asymmetric perspective and discretizing via quintiles indicate the usefulness of examining cases in all four complex outcomes via asking when does high, as well as low, KME indicate high PMP and when does high, as well as low, KME indicate low PMP rather than simply reporting a statistically significant, positive, variable-directional relationship.

The reality of causal asymmetry is the fourth complexity theory tenet. The causal conditions indicating cases having high outcome scores usually vary in content from the causal conditions indicating cases having low outcome scores. "Emergence" is the fifth complexity theory tenet in Table 2.1. The fifth tenet refers to the principle that a screen whereby a case has to achieve high scores (e.g., ≥0.90) for each of several conditions (a non-compensatory rule) is a tougher requirement than setting a screen whereby a case has to achieve a high score (e.g., ≥0.90) on the sum of several conditions (a compensatory rule).

3 Configurational Theory of Complex Antecedent Conditions Indicating High (Low) PMP

Figure 2.2 is a visual of the general asymmetric theory of antecedents and outcomes of projects high versus low in KME, in social capital, and in PMP. Figure 2.2 builds from the perspective that ongoing success in project management depends on choosing the right mix of new product strategy, organizational structure, and NPD processes (Barczak 1995; Henard and Szymanski 2001). This general theory focuses on predicting outcomes accurately with high consistency and not proposing and testing variable-directional hypotheses. Figure 2.2 is a visual of complex antecedent conditions for KME and social capital as well as KME and social capital as simple and within complex antecedent conditions indicating high versus low PMP outcomes. This section describes the eleven propositions appearing in Fig. 2.2.

P1: PMP ≤ KME The following asymmetric statement expresses P1. Cases with high in KME indicate high PMP consistently. The observed odds will be at least 4-to-1 is the "consistently high" metric that this study applies for testing all the propositions (e.g., if 10 cases are high in KME, 8 of these ten cases are high in PMP). "Crucial" and "critical" are words appearing often in the literature for describing knowledge management in the project management literature (e.g., Yang et al. 2012; Lindner and Wald 2011; Wiig 1997). This perspective raises at least two possibilities. First, cases high in KME are consistently high in PMP, that is, identifying projects achieving high KME is sufficient for identifying cases high in PMP. Second, while high KME is insufficient by itself to indicate cases consistently

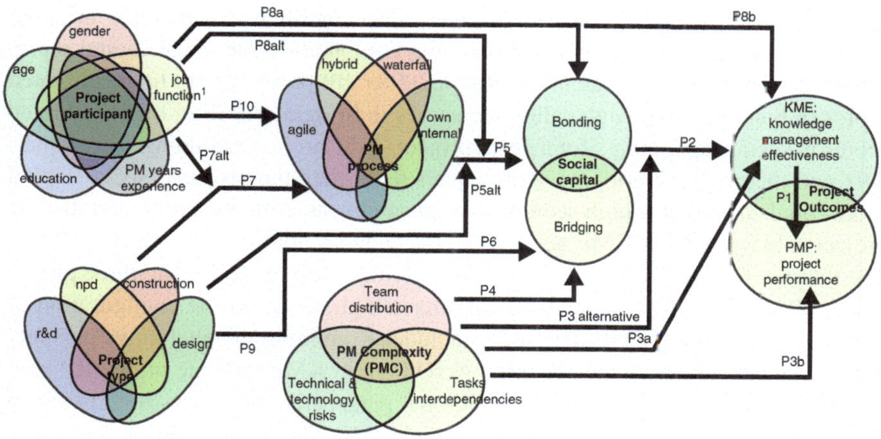

Fig. 2.2 The general theory of project management practices and performance outcomes. Note: [1]Job function = administration production/operations, R&D, sales/marketing, or others

high in PMP, the condition, high KME, is an ingredient in alternative models of complex antecedent conditions indicating high PMP consistently. The findings section explores both perspectives empirically. These findings provide strong support for the second perspective.

The present study also examines the impacts of the negation of KME (i.e., projects reported to have a comparatively low KME condition). The logical expectation receives support, that is, low KME is a condition in complex antecedent conditions frequently in indicating low PMP. However, the findings include one asymmetric model with high KME as one of the ingredients where the model indicates low PMP—thus, a useful view to adopt is that high KME alone is not a sufficient indicator of high PMP.

P2a: Social capital ≤ KME; P2b, social capital ≤ PMP Social capital includes "the norms and networks that create the necessary trust for people to cooperate to solve collective action problem" (Garrido 2014: 413). Colclough and Sitaraman (2005) and Gibbs et al. (2015) propose that the relationship that embeds in these social networks and in individual's positions within them provides people with a substantial amount of resources and power than they would normally hold (i.e., "linking" social capital), through access to resources such as information, knowledge, people, money, and power (i.e., "bridging"' social capital) and psychological support which builds trust and reciprocity ("bonding" social capital) (Teo and Lesomore 2017). Via bridging and bonding, social capital may enable and expand KME and PMP. Relatedly, the negation of social capital is likely to occur in projects having comparatively low KME and PMP.

Because the present study defines projects high in social capital to include high scores in separate metrics for both bonding and bridging mechanisms—the emergence tenet in complexity theory applies here, the study applies a strict operational definition of social capital as a complex antecedent condition (i.e., bonding•bridging = social capital). Consequently, the resulting computation for social capital represents a complex "fuzzy" statement (Ragin 2008a, b): projects with high calibrated scores in bonding AND high calibrated scores in bridging result in high calibrated scores in social capital (e.g., for a given project, if the project's bonding score equals 0.91 and the project's bridging score equals 0.88, then the projects' social capital score equals 0.88). This high score for social capital supports the prediction that the project achieves a high KME score (e.g., 0.91) and a high PMP score (e.g., 0.86). The findings section below includes results from testing the following three propositions. P2c: Cases high in social capital have high scores in the configuration of high KME AND high PMP. The study includes examining the alternative perspective—P2d: cases high in social capital AND high in KME indicate high PMP consistently. Also, the study examines additional perspective that cases high in social capital AND high in complexity indicate high PMP. These and additional propositions and findings provide a rich and deep understanding of the complex antecedent configurations indicating high (and modeled separately) low PMP.

P3: Low (~) project management complexity (PMC) is sufficient for indicating cases high in KME The present study measures three components of project management complexity (PMC): team distribution, technical and technological risks, and task interdependencies. Using a non-compensatory emergent calculation for PMC requiring all three components to have high scores for PMC to have a high score, a low score for PMC indicates a project comparatively simple to implement. Such implementation simplicity should provide for high KME. However, the findings do not support P3.

P4: High project management complexity (PMC) → social capital The rationale for P4 is that project managers working on highly complex cases are likely to recognize the need to generate high social capital to achieving high PMP. Though the three ingredients constituting PMC are ingredients in configurations indicating high PMP, this perspective that PMC → social capital does not receive support in the findings of the present study. The findings in the present study support the elimination of P4 from Fig. 2.2 and high value in examining the three ingredients in PMC (i.e., team distribution, technical and technology risks, and tasks interdependency) separately. Simply put, the ingredients in PMC are relevant in more than one recipe for projects resulting in high PMP.

P5: Projects implementing different project management processes nurture high social capital uniquely, but a given process model alone is insufficient for indicating high social capital The present study examines the impact of four PM process models: agile, waterfall, hybrid, and firm internal models. Organizations applying the agile process model organize project implementation in small interdisciplinary teams designed to complete a single project, rather than using a more formal hierarchical structure; the software keeps the team members both informed and connected (Bartleby 2018). Thus, cases implementing agile processes generate high social capital—the present study examines this proposition. The findings do not support this perspective (i.e., having cases high in the use of agile for PM processing was not an indicator of cases high in social capital). However, the study does find that the use of agile is an ingredient in models specifying specific categories of projects for indicating projects achieving both high KME and high PMP. Thus, agile does matter, but to learn how this tool matters, theory and research needs to go deeper than specifying agile as a simple main effect on social capital on project performance.

Waterfall process uses less iterative and flexible approaches than the agile model—as progress flows in largely one direction ("downwards" like a waterfall) through the phases of conception, initiation, analysis, design, construction, testing, deployment, and maintenance (Rodriquez and Williams 1998). The literature includes substantial criticisms of the waterfall model since the model's use typically operates in one direction (e.g., after needs are set, the setting is not revisited/revised based on downstream knowledge, and design changes). Rather than focusing on the relative worth of the agile versus waterfall models (i.e., instead of viewing project management processes to be competing against each other), the present study focuses on the more complex question of the contexts in which each process model is an ingredient in complex antecedent configuration models indicating high PMP (or models for low PMP). The study also constructs separate asymmetric models indicating when a specific PM process model is an ingredient in projects performing poorly compared to the performances of other projects.

A hybrid PM process is when activity is initially sequential and then iterative throughout the development process. "The hybrid model creates a disciplined timeline that allows for client alignment, but still flexible enough to iterate and refine. Due to the sudden shift from sequential to iterative, strong project management is needed, in order to keep the project's decision making on track" (Justin 2018). Firm "own internal" process models include highly modified versions with combinations of the characteristics of the other three PM process models. The respondents in the study were asked which of the four models best described the process in use in implementing their project; the respondents had the additional option to answer "Others" as well. Almost all respondents were able to select one of the first four process models to describe their projects' process models.

Before implementing the survey stage, the present study included completing long face-to-face interviews individually with ten members of project teams; these interviews indicate the possibility that members of some projects might be emphatic in describing their PM process to be highly unique and created internally by their firm/organization. Denison, Hart, and Kahn (1996) include related findings from 200 long open-ended interviews that members of some teams claim their PM processes to be unique. Thus, learning the presence, antecedents, and outcomes of this PM process uniqueness perspective prompted the inclusion of "own internal PM process" in the survey instrument.

Figure 2.2 in the present study includes the perspective that each of four PM processes may contribute to high social capital in different and sometimes in the same contexts—this perspective represents proposition 5 (P5). However, P5 comes with additional complexity: the impact of PM processes in indicating high social capital occurs only as ingredients in asymmetric models that include the matching of specific project types with PM processes. This complex perspective appears in Fig. 2.2 as an alternative to P5 (P5alt).

P5alt: Cases (i.e., projects) matching of specific project types and processes are high in social capital consistently Separately, mismatch combinations (i.e., configurations) of specific project types and processes are low in social capital consistently. Thus, the theory proposes that matching versus mismatching project types and processes is sufficient (though each model is unnecessary) for indicating cases high versus low in social capital, respectively.

For P5alt, a "computing with words" (CWW) (Zadeh 1996) is useful for stating specific propositions regarding product types and processes that should indicate high versus low social capital. The following are two examples. (1) A new product development (NPD) project using an agile process indicates high social capital. (2) A construction project using a waterfall process indicates low social capital. The present study examines these two statements empirically as well as the general propositions that a few recipes of combinations of project types and processes indicate high (low) social capital.

Though not appearing in Fig. 2.2, this match versus mismatch perspective for project types and process is extendable for indicating high and low PMP. Analogous to a classically trained symphony orchestra, various project type process combinations work well to result in superb outcomes while other combinations work to deliver bad performances. The present study examines and partially supports this perspective empirically.

P6: Project types alone do not indicate high (low) social capital consistently Empirically, P6 is achievable because knowledge of project type alone is likely to be insufficient for indicating high or low projects—also necessary is knowledge of project management process, and knowledge of specific project team member may need inclusion for indicating high social capital. However, the consistency index metric for predicting cases high in social capital based on project type alone is likely to be highest for NPD and lowest for construction projects. The proposal is that NPD projects require several mock-ups and revisions based on client reactions to prototypes—requiring several rounds of bonding and bridging stages—whereas construction projects require building adhering mostly to a given set of blueprints with few revisions once construction is underway and, consequently, low levels of bonding and bridging for such projects. In predicting high social capital, the consistency indexes for R&D and design projects should be higher than the consistency index for construction projects and similar to the consistency index for NPD projects. This study examines whether or not these prediction rankings are correct for the consistency levels in indicating high social capital for each of the four project types.

P7: Project type alone does not indicate the use of a specific project process Even though project type may indicate the relevance of the project management process, project type alone is insufficient information for consistently predicting the use of a specific project management process. In part, the selection/use of the project management process may depend on personal preferences and prior training of different members of the project team. Consequently, the combination of project type and information about project team participants is likely to be sufficient for indicating the project management process. The arrow for P7alt in Fig. 2.2 represents this alternative perspective. This study examines both P7 and P7alt empirically.

P8a: Knowledge of project team participant demographic profiles is sufficient for indicating projects high in social capital. P8b: Knowledge of project team participant demographic profiles is sufficient for indicating projects high in PMP Team members working on the projects differ frequently in training and expertise and vary by job function, gender, age, and years of project management experience—creating such requisite variety among team members enables bringing multiple skills to bear for achieving high PMP. However, while configurations of information describing specific team members may be a useful ingredient in complex antecedent conditions indicating high social capital, such information may be too meager to do so consistently. The present study examines these alternative perspectives empirically for social capital and PMP as the outcome conditions. P8alt in Fig. 2.2 represents the perspective that specific combination of project team member participant demographic profiles and project management process is sufficient for indicating high social capital consistently. The present study examines P8a, P8b, and P8alt empirically.

P9: Project type indicates high project complexity consistently Typically R&D projects are inherently more complex than construction projects. While the findings may indicate the R&D projects indicating high project complexity with a consistency greater than construction projects, if the bar for minimal consistency is set as a high index (e.g., 0.85), P9 is unlikely to receive support empirically. High consistency in predicting project complexity is likely to hold only from the perspective of R&D versus construction projects. The present study examines this perspective empirically.

P10: Project member demographic profiles are insufficient information to indicate project process consistently While project member participant may be a contributing ingredient in identifying a project management process, information about a project's participant is insufficient for accurately identifying the project process in use consistently. Even if an R&D-trained project participant prefers using the agile process method, her/his preference would likely be relevant for one project type sufficiently for indicating adoption of the agile process. The general tenet of complexity theory applies: information about a single ingredient (e.g., project participant description) alone is incapable of identifying a project management process consistently.

4 Method

This section describes the study's survey instrument, procedure for data collection, the data, and the data analyses steps.

4.1 Survey Instrument

The questions in the study include multiple-item scales and seven-point Likert scales to measure the constructs. Question options ranged from 1 ("strongly disagree") to 7 ("strongly agree"). Exploratory factor analysis using varimax rotation was used to extract the factors, and reliability of the constructs was done by calculating the Cronbach's alphas. The survey comprises of eight parts including bonding capital, bridging capital, project management performance, knowledge management effectiveness, team diversity, team distribution, task interdependence, technology, and technical risks and complexities. Prior studies informed the development of these items and they were contextualized into the domain of project management.

Bonding Capital The bonding capital scale included four items based on the work of Seashore (1954). The Cronbach's alpha was 0.85.

Bridging Capital The bridging capital items are based on the work of Ellison et al. (2007). In the factor analysis, one of the items cross-loaded on bonding capital and was subsequently removed from further analysis. The Cronbach's alpha was 0.89.

Project Management Performance The five items measuring PMP are based on the work of Malach-Pines et al. (2009) and adapted from Han and Hovav (2013). The Cronbach's alpha was 0.84.

Knowledge Management Effectiveness The three items measuring KME were adapted from the work of Song et al. (2007). The Cronbach's alpha for this scale equaled 0.81.

Project Complexity Project complexity is a multidimensional construct. Based on previous studies, it consists of task interdependence, team diversity, team distribution, and technical risks and complexities. Each second-order construct in this multidimensional construct is discussed below:

Technical Risk Complexity. This eight-item scale was adapted from Wallace et al. (2004). Cronbach's alpha for this scale equaled 0.85.

Task Interdependence. The five items measuring task interdependence were adapted from Van Der Vegt et al. (2000). One item did not load on the construct and was removed from further analysis. The Cronbach's alpha for this scale equaled 0.81.

Team Distribution. This four-item scale was adapted from Chudoba et al. (2005). The Cronbach's alpha for the revised scale equaled 0.87.

Team Diversity measures were adapted from the work of Campion et al. (1993). The three-item scale was not reliable (Cronbach's alpha was 0.64) and therefore subsequently removed from further analysis in this study.

4.2 Respondents

The present study includes an empirical examination of complexity tenets in the context of project participants, project management processes, social capital, KME, and PMP. The survey was completed by project managers from the USA using the survey questionnaire. The data collection for this study was done by the survey panel, Qualtrics, Inc., USA. Project managers were asked to complete the survey online. Because of the bias of the survey panel, several attention filters were included randomly in the questionnaire to increase the quality of the data. A total of 746 respondents completed the questionnaire; 197 failed an attention filter and were removed from the survey data, 246 were not project managers and were also removed from the survey, and 303 were project managers and successfully completed the survey. Of the 303 questionnaires, one was unusable due to incomplete responses and was subsequently removed from the analysis.

A total of 157 (52%) of the respondents were male, while 145 (48%) were female. In terms of age, 3% were between 18 and 24 years old, 43% between 25 and 34 years old, 38% between 35 and 44 years old, 12% between 45 and 54 years old, and 5% were 56 or older. Only one of the respondents reported having less than a year experience as a project manager (PM), 11.6% reported having 1–2 years of experience as a PM, 20.2% have 3–4 years of experience as a PM, 52.3% have 5–10 years of experience as a PM, 10.3% have 11–15 years of experience as a PM, and about 5.3% reported having more than 15 years of experience as a PM. A total of 124 respondents reported having a bachelor's degree (41%), 83 (27%) have a graduate degree (27%), 28 (9%) have an associate degree, 27 (9%) have some college, and 15 (5%) have a PhD degree. The processes used on the project are as follows. A total of 86 (28.4%) of the projects used the waterfall (traditional) method, 86 (28.4%) used the hybrid method, 52 (17.2%) used the agile, and 79 (26.2%) did not have an established project management process.

The primary functional areas of the respondents are as follows: 129 (42.7%) are in productions and operations, 53 (17.9%) are in research and development, 40 (13.2%) are in administration, 36 (11.9%) are in sales and marketing, and 43 (14.2%) are in human resources, purchasing, customer service, IT support, and finance and account. The breakdown of the type of project is as follows: 63 (20.9%) are construction projects, 49 (16.2%) are design projects, 40 (13.2%) are R&D projects, 37 (12.3%) are new product development projects, and the remaining 113 (37.4%) are maintenance, administrative, computer software, strategic, architectural development, applications, and infrastructure projects.

4.3 Analysis

The current study employs a set-theoretic approach based on fuzzy set QCA that allows for a detailed analysis of how causal conditions contribute to an outcome of interest. Fuzzy set QCA is suitable for analyzing causal processes because it provides a configurational understanding of how causes combine to lead to different outcomes (Fiss 2007; Ragin 2000, 2008a, b). The basic tenet underlying QCA is that cases are best understood as configurations of attributes and that a comparison of cases can allow a researcher to sort out attributes that are unrelated to the outcome of interest (Fiss 2011). Set-theoretic analysis examines causal patterns by focusing on the subset relationship.

> For instance, to explain what configurations lead to high performance, it examines members of the set of "high-performing" organizations and then identifies the combinations of attributes associated with the outcome of interest (high performance) using Boolean algebra and algorithms that allow logical reduction of numerous, complex causal conditions into a reduced set of configurations that lead to the outcome (Fiss 2011: 402).

Fuzzy-set theory (Zadeh 1965, 1996, 2010) enables expressing a case's degree of membership in a set in fine gradations ranging from full nonmembership (0.05 and below) to full membership (0.95 and above). The analysis requires calibrating the original value for a variable for a case against a range of membership scores that represent the accumulated knowledge about a set. The membership scores that represent the set are estimated by examining information on the entire set (Ragin 2008a) to identify the threshold for full nonmembership (set at 0.05), the point of "maximum membership ambiguity" (set at 0.50), and the threshold of full membership (set at 0.95). Calibrating original values into fuzzy-set scores transforms ordinal or interval scales into degree of membership in the target set. Ragin (2008b: 183) refers to this as "a truth value" that reflects each case's position in an established set, which allows comparison of each case against a known external standard and enables the identification of fine gradations in the degree of membership of a set. In the current study, all antecedent conditions and the outcome conditions were calibrated based on records for the full set for each condition. The analysis includes the median value as the point of maximum ambiguity (0.50); full membership (0.95) was equal to the value immediately below the highest 14% of all available projects; full nonmembership was equal to the value immediately above the lowest 14% of all available projects. The calibration subroutine in the software program (available at fsQCA.com) computed the scores for all nations in the study. With no appreciable changes resulting to the findings, the present study included performing an alternative calibration procedure recommended by Ragin (2008a, b) whereby the 0.05 value was set at the 14th percentile level and the 0.95 score was set at the 86th percentile value. For all conditions, three slightly different sets of breakpoints were tested to confirm that the findings did not vary in calibrating either full- or nonmembership points. The calibration subroutine in the software, fsQCA.com, applies a logarithmic function in calibrating all other original values into membership scores.

4.4 Consistency and Coverage

Consistency is a measure of the degree to which a specific recipe is consistent with the outcome. A high consistency score indicates that nearly all cases with a high membership score in the recipe of conditions also have a high score in the outcome condition. Consistency occurs when a case's level of membership in the causal recipe is less than or equal to its level of membership in the outcome (i.e., the causal recipe is a subset of the outcome) (Ragin 2008a, provides explicit examples for computing consistency and coverage). The fsQCA software estimates consistency ($Xi \leq Yi$), by dividing the sum of the minimum of Xi or Yi by the sum of Xi. Consistency $(\mathbf{Xi} \leq \mathbf{Yi}) = \Sigma\ \{\min\ (\mathbf{Xi}, \mathbf{Yi})\}/\Sigma\ (\mathbf{Xi})$, where Xi is case i's membership score in set X; Yi is case i's membership score in the outcome condition, Y. This approach penalizes consistency scores for occurrences when Xi exceeds Yi in proportion to the size of the inconsistency. Calculation of consistency is definable as performing "somewhat precise outcome testing (SPOT)" (Woodside 2017). SPOT is an alternative summary metric to symmetric test's null hypothesis statistical tests (NHST) (Woodside 2017).

The coverage index in fsQCA assesses the degree to which a causal configuration accounts for instances of an outcome. Coverage should only be considered after consistency is established. Set theory accommodates the idea that there are alternate causal recipes that lead to the outcome. QCA compares alternate configurations by determining the proportion of cases that are consistent with a particular configuration (Ragin 2008a). If a configuration accounts for more cases, then it is more informative. To calculate coverage, the fsQCA software calculates the proportion of cases where degree of membership in the configuration is less than the membership in the outcome, but this time divides it by the total score for the outcome condition. This coverage calculation estimates the level to which the configuration (the subset) covers the outcome (the larger set) (Ragin 2008a). Coverage $(\mathbf{Xi} \leq \mathbf{Yi}) = \Sigma\ \{\min\ (\mathbf{Xi}, \mathbf{Yi})\}/\Sigma\ (\mathbf{Yi})$.

Ragin (2008a) provides a full description of QCA theory and method. Based on the equifinality tenet in complexity theory, the coverage of models with high consistencies are not expected to provide unique high coverages of the number of cases with high outcome membership scores. For the models with high consistencies reported in the findings, coverages are above 0.05. The study here focuses on theory-based ex ante rather that post hoc predictions that are functionally accurate—similar to the claim of a poker player of having a system that works for winning a poker hand if a certain two cards appear in the draw but providing no information if other cards are drawn.

5 Findings

The use of Boolean algebra in this study provides the complex configurations of conditions (i.e., asymmetric models) that are useful for forecasting high project management performance (PMP) accurately. Separate asymmetric models can be created for forecasting low PMP accurately. Fuzzy-set qualitative comparative analysis (fsQCA) was used for modeling both positive and negative paths to high scores in an outcome variable. Table 2.2 demonstrates the occurrence of both negative and positive cases for the relationships between KME and PMP. What combination of antecedents' conditions includes negative KME cases that indicate high PMP? What combination of antecedent conditions includes positive KME cases that indicate high scores in the negation of PMP? The next section describes the fuzzy-set QCA combinations. QCA studies cases as configurations of causes and situations rather than treating each independent variable in the statistical analysis as distinct and separate which is what is done in conventional quantitative methods.

5.1 Detailed Findings

P1 receives partial support—high KME is sufficient for indicating high PMP.

The findings in Table 2.3 and Fig. 2.3 partially support P1. The consistency index is 0.779 and the coverage is 0.723. These findings support the perspective that high KME alone is insufficient for predicting high PMP. The asymmetric claim earlier was that the odds would be greater than 4-to-1 for high KME cases indicating high PMP outcomes. The findings include observed odds equal to 2-to-1 and not 4-to-1 for high KME cases indicating the high PMP outcome. The project cases low in KME and high in PMP support the perspective that additional asymmetric models are needed that may include the negation of KME and additional conditions for indicating high PMP consistently.

P2a receives partial support—high social capital is modestly sufficient for indicating high KME.

The findings in Table 2.3 and Fig. 2.4 partially support P2a with a consistency index of 0.808 and coverage of 0.66. Bonding capital by itself and bridging capital by itself do not predict KME consistently.

P2b receives partial support—high social capital is modestly sufficient for indicating high PMP.

The findings in Fig. 2.4 support P2b with a consistency index of 0.86 and coverage of 0.65. Bonding capital by itself and bridging capital by itself do not predict cases with high PMP. Bonding and bridging capital have to be both present to predict cases high in PMP. The XY plot from testing P2b (Fig. 2.5) indicates that the odds are greater than 4-to-1 for high social capital cases indicating high PMP outcomes.

P2c does not receive support. Cases high in social capital are not consistently high in the joint outcome for KMP•PMP. The consistency index for this model is equal to 0.73 with a coverage equal to 0.76.

P2d receives support—social capital and KME are sufficient for indicating cases high in PMP.

The findings in Table 2.3 and Fig. 2.6 support P2d. The consistency index equals 0.90 and the coverage index equals 0.55 for this model. The high consistency

Table 2.3 General theory propositions and findings: mainly testing simple antecedents statements as necessary conditions

Propositions	Consistency (C1)	Coverage (C2)	Pi Supported?
P1 High KME is sufficient for indicating (→) high project management performance (PMP)	0.78	0.72	Partially
P2a High social capital (SC) is sufficient for indicating (→) high KME,, bonding•bridging → KME	0.81	0.66	Partially
P2b High social capital is sufficient for indicating (→) high PMP, that is, bonding•bridging → PMP	0.86	0.65	Yes
P2c High social capital is sufficient for indicating (→) high KME•PMP	0.73	0.76	No
P2d: Alternate perspective: social capital AND KME (SC•KME) indicate cases high in PMP	0.90	0.55	Yes
P2e: Alternative perspective: social capital AND complexity indicate cases high in PMP	0.91	0.35	Yes
P3a ~PM complexity is sufficient for indicting (→) high KME	0.50	0.31	No
P3b ~PM complexity is sufficient for indicating (→) high PMP	0.65	0.32	No
P3c ~PMC → KME•PMP (low complexity indicates high overall performance)	0.40	0.32	No
New alternative PMC perspectives:			
P3aa High PMC is sufficient for indicating (→) KME	0.81	0.46	Partially
P3bb High PMC is sufficient for → PMP	0.81	0.42	Partially
P3cc High PMC is sufficient for → KME•PMP	0.71	0.52	No
P4 High PM complexity → social capital	0.74	0.51	No
P5 Project management process → social capital	0.00	0.00	No
P6 Project type → social capital	0.00	0.00	No
P7 Project type → project process	0.00	0.00	No
P8a Project participant demographic profiles → social capital	0.85	0.10	Yes
P8b Project participant demographic profiles AND project management process → social capital	0.89	0.11	Yes
P9 Project type → project management complexity	0.00	0.00	No
P10 Project participant → PM process	0.00	0.00	No

Note: Ex ante set requirement for high consistency ≥0.82. Key: *KME* knowledge management complexity, *PMC* project management complexity, *PMP* project management performance; "→" "is sufficient for indicating"; "•"; "~" indicates negation (i.e., 1 minus the causal condition)

supports the conclusion that the combination of high social capital (bonding and bridging) in combination with high KME is sufficient for predicting cases having high PMP consistently. The findings in Fig. 2.6 show that the odds are greater than 4-to-1 for projects having high scores for the complex antecedent condition indicating high PMP scores.

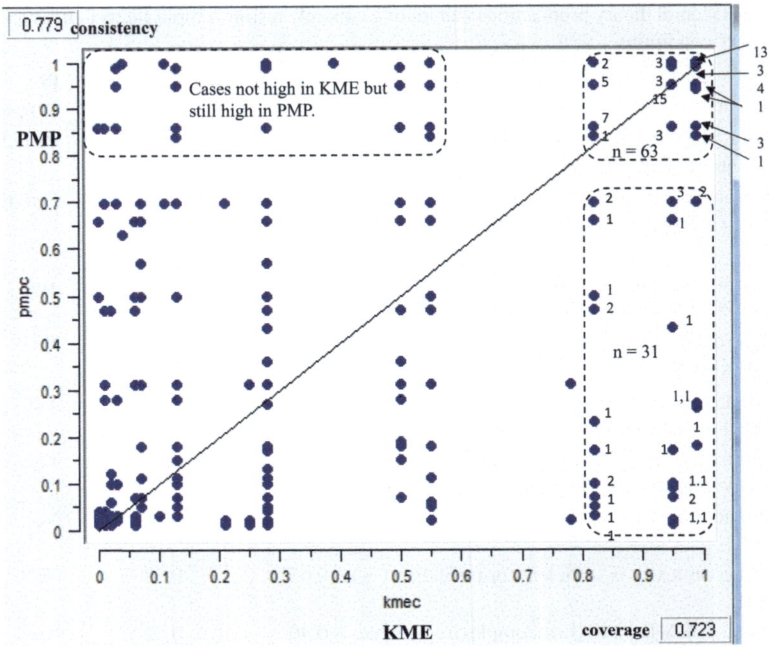

Fig. 2.3 Findings for model for P1: KME ≤ PMP. Knowledge management effectiveness (KME) indicating project management performance (PMP)

Note: The following is an interpretative guide for reading the XY plots in the study

Numbers in Fig. 2.2 are the numbers of cases per dot

This model is an asymmetric analysis with the objective of constructing a model that is sufficient (but not necessary) for indicating cases with high membership scores in the outcome condition

The asymmetric model, KME ≤ PMP (i.e., KME → PMP), is useful! Cases high in knowledge management effectiveness (KME) alone are a reasonable useful indicator of high project management performance (PMP). The odds are 2 to 1 (63/31), indicating that cases high in KME are high in PMP

However, high KME does not predict all cases high in PMP. Thus, high KME is not necessary for high PMP. But the model demonstrates sufficiency for indicating cases high in PMP

P2e receives support—social capital AND project complexity are sufficient for indicating cases high in PMP. The findings in Table 2.3 support P2e. Model 2 in Table 2.3 represents cases low in technical risks and low in team distribution but high in KME, and bonding capital can predict cases high in PMP. The findings indicate that impact of project complexity, social capital, and KME is not always positive or negative. PMP is too complex to be captured by main effects and net effects.

P3a does not receive support. Cases reported to be low in PMC alone are a poor indication of cases high in KME. Because the findings below refute P3 and the abductive insight that cases high in PMC may cause team members to recognize the need for nurturing high KME, the study examines and finds support for this alternative perspective: cases high in PMC are sufficient for indicating cases high in KME, P3aa: PMC → KME. The consistency index for P3aa equals 0.81.

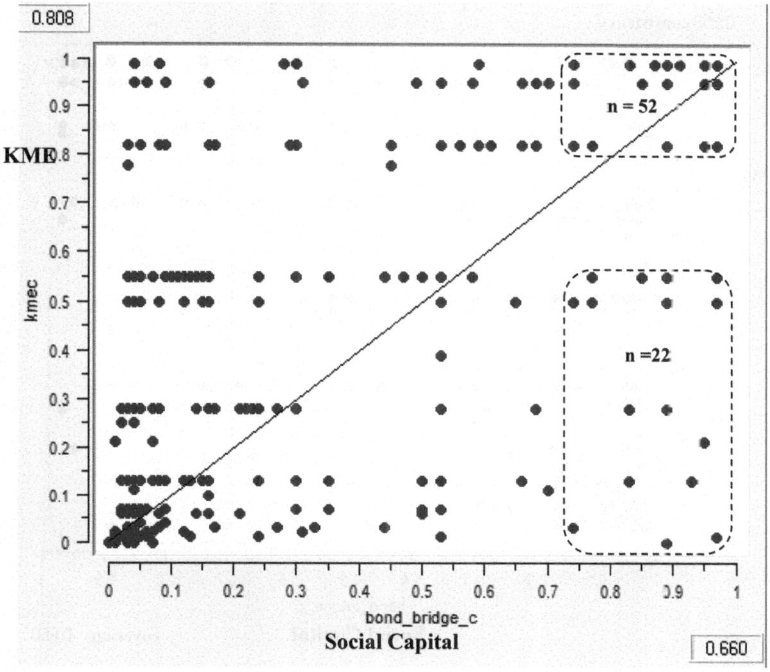

Fig. 2.4 Model for P2a: Social capital ≤ KME. Social capital (bonding and bridging) indicating knowledge management effectiveness (KME)

The findings for PMC and PMP are similar as the findings for PMC and KME. The negation of PMC does not indicate cases high in PMP. P3b does not receive support. The reverse forecast receives support. Most of the cases high in PMC are high in PMP. The data analysis indicates that the odds are 6-to-1, suggesting that cases high in PMC are high in PMP. (This paper does not include the XY plot for this finding due to space limitations.) Appearing in Table 2.3, P3bb is the simple statement that PMC → PMP; the consistency index for the cases for this model equals 0.81 and the coverage index is equal to 0.42. Forecasting the complex outcome of the combination of KMC AND PMP indicates that neither PMC nor the negation of PMC is sufficient for doing so. Details of findings from testing these two simple models appear in Table 2.3 (P3c and P3cc).

P4 does not receive support: cases high in PMC are insufficient for indicating high social capital. The consistency is 0.74 with a coverage of 0.51 as Table 2.3 shows.

P5 does not receive support: cases high in project management process do not indicate cases high in social capital consistently. The findings in Table 2.3 refute P5 with a consistency of 0.00 and coverage of 0.00. This finding does not support the perspective that knowledge of PM process by itself is irrelevant for indicating cases high in social capital. However, specific PM process versus others may be one among a few ingredients in models indicating high (low) social capital. The findings from testing P4, P5, P6, and P7 support the first complexity tenet that a single antecedent condition is insufficient usually for indicating cases with high scores in an

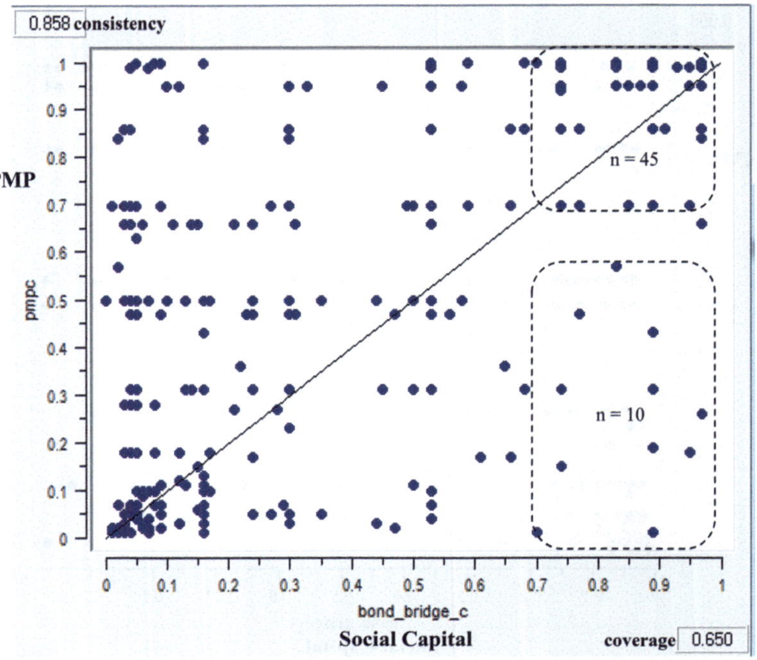

Fig. 2.5 Model for P2b: Social capital ≤ PMP. Social capital (bonding and bridging) indicating project management performance (PMP)

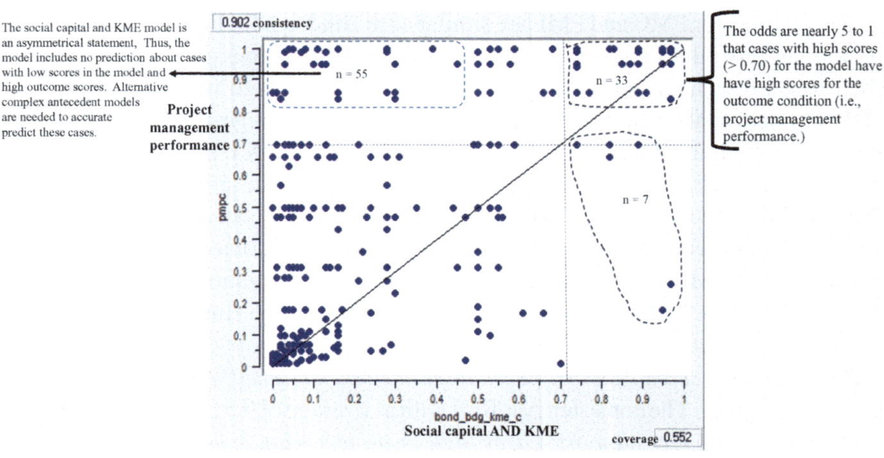

Fig. 2.6 Findings for alternative asymmetric perspective of cases high in social capital AND high KME indicating cases high in PMP (project management performance)

outcome condition. This perspective receives support generally even when the correlation between the single antecedent and outcome condition is very high (e.g., $r = 0.70, p < .001$). Thus, if a researcher focuses on predicting outcomes accurately rather than showing that one or more relationships do not have a zero correlation, the researcher needs to construct and test theories of recipes (cf. Ordanini, Parasuraman, & Rubera, 2014).

P6 does not receive support: cases high in project type alone do not indicate cases high in social capital consistently. The consistency is 0.00 with a coverage of 0.00 as Table 2.3 shows.

P7 does not receive support: cases high in project type alone is insufficient for indicating cases of project management process. For the specific findings in testing P7, Table 2.3 reports that the consistency index equals 0.00 and the coverage index equals 0.00 (Table 2.4).

P8a receives support—project-participant demographic configurations are sufficient for indicating cases high in social capital. The findings in Table 2.3 provide details from testing P8a. P8a predicts that high scores for specific demographic configurations indicate high scores for the complex outcome of bonding AND bridging (i.e., social capital). Table 2.5a provides details for the findings for P8a. For instance, model 1 on Table 2.5a indicates that cases having the following PM membership profile have high social capital: older male managers with high years of experience as a project manager whose functional working area is sales and marketing and with high or low education. Examining Table 2.5a indicates that the demographic feature can be positive or negative for different recipes with the outcome of high social capital occurring (i.e., the complexity theory equifinality tenet receives support)—with one exception. Manager with high numbers of years of experience is a feature in all three configurations in Table 2.5b.

Table 2.4 Alternative models that work well: configurational models indicating high PMP and models indicating the negation of PMP consistently

Testing models indicating high project management performance (pmp) = f(task_inter_c, t_t_risks_c, team_d_c, kmec, bridge_c, bond_c)
INTERMEDIATE SOLUTION: frequency cutoff: 2.00; consistency cutoff: 0.86.

Models predicting high PMP	coverage	consistency
1. kmec• bridge_c• bond_c	0.55	0.90
2. ~t_t_risks_c• ~team_d_c• kmec• bond_c	0.27	0.85
3. ~t_t_risks_c• ~team_d_c• bridge_c• bond_c	0.29	0.86
4. task_inter_c• t_t_risks_c• kmec• bridge_c	0.41	0.91
5. t_t_risks_c• team_d_c• bridge_c• bond_c	0.38	0.90
6. task_inter_c• t_t_risks_c• ~team_d_c• ~bridge_c• bond_c	0.16	0.86

Solution coverage: 0.71; solution consistency: 0.85.

Notes.
- Here is an example of reading these "computing with words" models.
- For model 1: Cases high in "knowledge management effectiveness" AND high in social capital (i.e., both bridging and bonding) are high in project management performance (PMP) consistently (•.90 index).
- The equifinality tenet applies, multiple routes occur for reaching high PM: six unique models indicate the same outcome, that is, high project management performance. Note that high bonding is a feature in five of the six models; the negation of bridging appears in model 6. Both high and low technical and technology risks are features in different models.

Testing models indicating low project management performance (~pmp) = f(bridge_c, bond_c, task_inter_c, t_t_risks_c, team_d_c, kmec)
INTERMEDIATE SOLUTION: frequency cutoff: 2.00; consistency cutoff: 0.85

Models	coverage	consistency
1. ~bridge_c• ~task_inter_c• ~team_d_c• ~kmec	0.44	0.88
2. ~bridge_c• ~bond2_c• task_inter_c• ~kmec	0.32	0.84
3. ~bridge_c• ~task_inter_c• t_t_risks_c• ~kmec	0.30	0.90
4. ~bridge_c• ~bond2_c• ~task_inter_c• kmec	0.24	0.82
5. ~bond2_c• ~task_inter_c• ~t_t_risks_c• ~team_d_c• ~kmec	0.36	0.88
6. ~bridge_c• task_inter_c• ~t_t_risks_c• team_d_c• ~kmec	0.21	0.89
7. ~bridge_c• ~bond2_c• ~t_t_risks_c• team_d_c• kmec	0.19	0.84

Solution coverage: 0.67; solution consistency: 0.83

Notes.
- The negation of bridging and/or bonding (i.e., social capital components) appear in all models indicating low PMP. Thus low social capital is a generalizable feature in models indicating ~PMP—even though low social capital but itself is insufficient for indicating ~PMP..
- Both high and low knowledge management effectiveness appear in different models indicating ~PMP consistently. The finding illustrates the complexity theory tenet that the same feature has both a positive and negative association for a given outcome—which depends on additional features in different complex antecedent conditions.

Key: *PMP* high project management performance, *~PMP* low project management performance, *KMEC* calibrated knowledge management effectiveness, *bridge_c* calibrated bridging, *bond_c* calibrated bonding, *t_t_risks_c* calibrated technical and technology risks, *team_d_c* team distribution, *task_inter_c* calibrated task interdependencies

Table 2.5a P8a receives support: project manager demographic profiles accurately predicting high social capital

Model: bond_bridge_c = f(mgr_yr_exp_c, age_c, edu_c, r_d, admin, prodops, salesmtg, gender_c)
Intermediate solution: frequency cutoff, 1.00; consistency cutoff, 0.87

Models	Coverage	Consistency
1. mgr_yr_exp_c•~age_c•~r_d•~admin•~prodops•salesmtg•gender_c	0.04	0.94
2. mgr_yr_exp_c•age_c•~edu_c•~r_d•~admin•~prodops•~salesmtg•~gender_c	0.03	0.87
3. mgr_yr_exp_c•age_c•~edu_c•r_d•~admin•~prodops•~salesmtg•gender_c	0.07	0.88
Solution coverage: 0.10; solution consistency: 0.86		

Note: Example of reading models ("computing with words"): model 1 states that male managers with high years of experience working in sales and marketing, with high or low education, indicate high social capital

Table 2.5b Project manager demographic profiles accurately indicate cases having high project management performance

Model: pmp = f(mgr_yr_exp_c, age_c, edu_c, r_d, admin, prodops, salesmtg, gender_c)
Intermediate solution: frequency cutoff, 2.00; consistency cutoff, 0.86

Models	Coverage	Consistency
1. ~mgr_yr_exp_c•~age_c•edu_c•~r_d•~prodops•~salesmtg•gender_c	0.08	0.85
2. ~age_c•edu_c•~r_d•~admin•~prodops•salesmtg•gender_c	0.05	0.95
3. mgr_yr_exp_c•age_c•~edu_c•~r_d•admin•~prodops•~salesmtg•~gender_c	0.04	0.88
4. ~mgr_yr_exp_c•age_c•edu_c•~r_d•admin•~prodops•~salesmtg•~gender_c	0.05	0.90
5. mgr_yr_exp_c•age_c•~edu_c•r_d•~admin•~prodops•~salesmtg•gender_c	0.05	0.87
6. mgr_yr_exp_c•~age_c•edu_c•r_d•~admin•~prodops•~salesmtg•gender_c	0.05	0.91
Solution coverage, 0.21; solution consistency, 0.85		

Example reading, model 1: young, male, project manager with few years of experience, high education level, not working in production/operations nor in sales/marketing nor in R&D

Table 2.5c P8c receives support: PM process and project manager demographic profiles accurately indicating cases having high scores in the complex outcome of high KME AND high PMP

Testing P8b: Model: kme_pmp_c = f(agile_c, r_d, admin, prodops, salesmtg, gender_c, mgr_yr_exp_c, age_c, edu_c)
Intermediate solution: frequency cutoff, 1.00; consistency cutoff, 0.86

Models	Coverage	Consistency
1. agile_c•~r_d•~admin•prodops•~salesmtg•~gender_c•~age_c•edu_c	0.05	0.93
2. agile_c•~r_d•~admin•prodops•~salesmtg•mgr_yr_exp_c•~age_c•edu_c	0.08	0.90
3. ~agile_c•r_d•~admin•~prodops•~salesmtg•gender_c•mgr_yr_exp_c•~age_c•~edu_c	0.06	0.86
4. agile_c•~r_d•~admin•~prodops•salesmtg•~gender_c•~mgr_yr_exp_c•~age_c•edu_c	0.03	0.94
5. agile_c•~r_d•admin•~prodops•~salesmtg•gender_c•~mgr_yr_exp_c•~age_c•edu_c	0.04	0.90
6. agile_c•r_d•~admin•~prodops•~salesmtg•~gender_c•mgr_yr_exp_c•~age_c•edu_c	0.03	0.98
7. ~agile_c•~r_d•~admin•~prodops•salesmtg•gender_c•mgr_yr_exp_c•~age_c•edu_c	0.04	0.90

Solution coverage, 0.17; solution consistency, 0.86

P8b receives support as well: PM participant demographic profiles indicate cases high in PMP. Table 2.5b provides details. The same high or low demographic feature does not appear consistently for all six models in Table 2.5b. At the same time, specific deep (i.e., configurational) descriptions of project managers do indicate high PMP.

P8b receives support—project participant demographics AND project management process in combination are sufficient for indicating cases for the complex outcome high in KME AND high PMP. The findings in Table 2.3 support P8b with an overall solution consistency of 0.85 and coverage of 0.08. The demographics of the project participants AND project management process are sufficient for predicting high scores in the complex outcome of KME and PMP. For example, model 1 in Table 2.5c predicts that cases with the following PM process and demographic profile have high KME AND PMP: a young female is highly educated and works in production operations, the project is not R&D and uses the agile project process, and project management experience is irrelevant in this model.

P9 does not receive support: project type alone is insufficient for indicating cases high in PMC. The findings on Table 2.3 refute P9 with findings including a consistency of 0.00 and coverage of 0.00

P10 receives support: project participant demographic profiles alone are insufficient for indicating project management process. As expected the findings in Table 2.3 support the perspective that project participant demographic profiles alone do not indicate PM process in use (consistency of 0.00 and coverage of 0.00).

The present study also includes splitting the sample into two equal-sized subsamples (n = 151 for each) and performing predictive cross-validation checks. The specific models predicted by the theory and indicated by the findings in the first sample were tested on the second sample and vice versa. The two sets of analysis provide partial evidence of the generality of the findings. However, additional studies are necessary for confirming the general theory and the specific findings in this paper.

6 Discussion, Contributions, and Limitations

By recognizing the inherent complexity in achieving high (as well as low) project management performance (PMP), in the context of project management, the study here offers workable theoretical and empirical solutions to "the crucial problem in strategic management is firm heterogeneity—why firms adopt different strategies and structures, why heterogeneity persists, and why competitors perform differently" (Powell et al. 2011, p. 1370). The solutions include embracing Weick's (2007, p. 17) call for richness in the study of project management (PM): "Simple accounts mean that you are not paying attention." Applying and confirming complexity theory tenets in the present study enriches theory construction and empirical testing to represent the inherent richness and complexity occurring in PM. Testing the consistency of predicting outcomes with algorithms (i.e., "computing with words;"

Zadeh 1996, 2010) rather than testing the existence of variable relationships via null hypothesis significance tests (NHST) enables shifting to good science practice and away from bad science practice (cf. Hubbard 2015; Wasserstein and Lazer 2017; Ziliak and McCloskey 2008).

Answering the two questions at the beginning of this paper, both the general theory and findings in the present study indicate that high KME alone does not indicate high PMP consistently. Most cases achieving high social capital achieve high PMP—consistency equals 0.86 and coverage equals 0.66. However, projects achieving the configuration of high KME and high social capital do achieve high PMP consistently—the first model in Table 2.4 indicates that this case-based forecasting model provides a consistency index equal to 0.90 and a coverage equal to 0.55. This coverage is quite high, but the theory and findings support the occurrence of additional models indicating high PMP. However, all models include either high KME and/or high scores in one or both of the two features of social capital (Table 2.4 offers the details).

It is given that projects high in KME and/or social capital indicate projects high in PMP. What case configurations indicate high KME and high social capital? The general model and findings support the perspectives that alternative configurations that include project management participants, processes, project type, and/or project complexity impact both KME and social capital. The theory and findings extend prior work on asymmetric modeling in project management by Ordanini, Parasuraman, and Rubera (2014); individual project management attributes have complex trade-off effects that only certain combinations of attributes act as sufficient conditions for high PMP.

The general theory here and findings support the perspective that the study of low project management performance (~PMP) requires models with features and configurations separate and unique from models of high PMP. This perspective is suggestive for scholars working on topics relating to the highly reliable organization (HRO), for example, Weick (1987), LaPorte, and Consolini (1991). and Weick, Sutcliffe, and Obstfeld, D. (2008). PMP outcomes include failure, mediocrity, and success. The evidence in the present study provides indication of how low PMP occurs and that forecasting projects resulting in low PMP is possible.

The view that low complexity associates with high KME is overly simplistic. The findings in the present study are suggestive that projects having high complexity attributes may frequently associate with high KME and high PMP. Organizations are likely to gear up to achieve high KME and high PMP when projects include high complexity. High project complexity by itself is not a consistent indicator neither for high nor low PMP.

The general theory and findings in the present study serve to indicate that models of high and low PMP should include modeling the impacts of PM participants' demographic profiles, project types, and project processes. Research in the PM literature frequently reports classifications and distributions for these features in the data collected for a wide range of projects but ignores these features in modeling. The general theory supports the perspective that these PM features can have direct and/or indirect impacts on projects' social capital, KEM, and performance.

From the practical perspective of considering how to nurture PMP, the theory and findings here support the perspective of examining whether or not project team members believe, and can provide evidence, that their project has achieved (is achieving) high social capital and high KME and taking steps to achieve both. Achieving high social capital and high KME explicitly as intermediate outcomes are worthy pursuits that many ongoing projects may ignore attempting to perform. Measuring explicitly the state of social capital and KME at early, mid, and later stages of project enactments is the main practical recommendation from the present study.

Limitations of the present study include the need for examining the general theory in additional studies on PMP. Also, additional studies are necessary that include asymmetric modeling of data that includes interviewing two or more team members working on the same project. Though rarely appearing in the extant literature, the possibility exists of examining projects where project team members' responses are in high-, medium-, and low-level agreement and to build separate asymmetric models for each of these three sets of projects (e.g., Ching, Chang, & Li, 2012). Also, the need exists to not rely solely on project managers' responses to surveys but to directly observe their behavior as well in field studies and subsequently to construct asymmetric forecasting models (e.g., Gladwin 1980). Consequently, while the general theory and findings in the present study are encouraging, they are limited by both depth and scope of coverage.

7 Conclusion

The present study supports two primary conclusions. First, while achieving high PMP is not necessary, achieving high social capital and high KME does frequently associate with cases achieving high PMP. Also, low social capital and/or low KME indicates cases achieving low PMP. The orderly, frequent, and consistent occurrences of high social capital and high KME in models indicating cases high in PMP coupled with the frequent occurrences of low social capital and low KME indicating low PMP support the perspective that advancing theory of PMP benefits from the study of both project social capital and KME.

Second, the theory and findings in the present study support the view that applying complexity theory tenets and using asymmetric modeling methods are viable solutions to overcoming the limitations of symmetric modeling using MRA/SEM with null hypothesis significance tests (NHST). In his Nobel Prize acceptance presentation, Simon (1978) provides a telling explanation that helps to understand the current widespread use of these "corrupt research" (Hubbard, 2015) practices.

> There is a saying in politics that 'you can't beat something with nothing'. You can't defeat a measure or a candidate simply by pointing to defects and inadequacies. You must offer an alternative. The same principle applies to scientific theory. Once a theory is well entrenched, it will survive many assaults of empirical evidence that purports to refute it unless an alternative theory, consistent with the evidence, stands ready to replace it (Simon 1978, p. 366).

References

Armstrong, J. S. (2012). Illusions in regression analysis. *International Journal of Forecasting, 28*, 689–694.

Banihashemi, S., Hosseini, M. R., Golizadeh, H., & Sankaran, S. (2017). Critical success factors (CSFs) for integration of sustainability into construction project management practices in developing countries. *International Journal of Project Management, 35*, 1103–1119.

Barczak, G. (1995). New product strategy, structure, process, and performance in the telecommunications industry. *Journal of Product Innovation Management, 12*, 224–234.

Bartleby. (2018). For richer, not poorer. *The Economist, 429*, 60, Economist.com/blogs/barleby.

Björk, J., & Magnusson, M. (2009). Where do good innovation ideas come from? Exploring the influence of network connectivity on innovation idea quality. *Journal Product Innovation Management, 26*, 662–670.

Campion, M.A., Medsker, G.J. & Higgs, A.C. (1993). Relations between work group characteristics and effectiveness: implications for designing effective work groups. *Personnel Psychology, 46*, 823–50.

Chereshnyuk, O., Panasyuk, V., Sachenko, S., Banasik, A., &, Golyash, I. (2017). *Fuzzy-multiple approach in choosing the optimal term for implementing the innovative project*. Proceedings, the 9th IEEE international conference on intelligent data acquisition and advanced computing systems: Technology and applications 21–23 September, 2017, Bucharest, Romania, pp. 532–536.

Colclough, G., & Sitaraman, B. (2005). Community and social capital: What is the difference? *Sociological Inquiry., 75*(4), 474–496.

Coleman, J. S. (1988). Social capital in the creation of human capital. *American Journal of Sociology, 94*(Supplement), S95–S120.

Denison, D. R., Hart, S. L., & Kahn, J. A. (1996). From chimneys to cross-functional teams: Developing and validating a diagnostic model. *Academy of Management Journal, 39*, 1005–1023.

Earl, M. (2001). Knowledge management strategies: Toward a taxonomy. *Journal of Management Information Systems, 18*, 215–233.

Ellison, N. B., Seinfeld, C., & Lampe, C. (2007). The benefits of Facebook "friends": Social capital and college students' use of online social network sites. *Journal of Computer-Mediated Communication, 12*(4), 1143–1168.

Fiss, P. C. (2007). A set-theoretic approach to organizational configurations. *Academy of Management Review, 32*, 1180–1198.

Fiss, P. C. (2011). Building better casual theories: A fuzzy set approach to typologies in organizational research. *Academy of Management Journal, 54*, 393–420.

Garrido, S. (2014). Plenty of trust, not too much cooperation: Social capital and collective action in early twentieth century eastern Spain. *European Review of Economic. History, 18*, 413–432.

Gibbs, A., Campbell, C., Akintola, O., & Colvin, C. (2015). Social contexts and building social capital for collective action: Three case studies of volunteers in the context of HIV and AIDS in South Africa. *Journal of Community Applied Social Psychology, 25*, 110–122.

Gigerenzer, G. (1991). From tools to theories: A heuristic of discovery in cognitive psychology. *Psychological Review, 98*, 254–267.

Gladwin, C. H. (1980). A theory of real-life choice: Applications to agricultural decisions. In P. Bartlett (Ed.), *Agricultural decision making: Anthropological contributions to rural development* (pp. 45–85). New York: Academic Press.

Han, J., & Hovav, A. (2013). To bridge or to bond? Diverse social connections in an IS project team. *International Journal of Project Management, 31*(3), 378–390.

Henard, D. H., & Szymanski, D. M. (2001). Why some new products are more successful than others. *Journal of Marketing Research, 38*, 362–375.

Hubbard, R. (2015). *Corrupt research: The case for reconceptualizing empirical management and social science*. Los Angeles: Sage.

Justin, A. (2018). Project framework comparisons: Agile vs. Waterfall vs. Hybrid vs. Lean. Downloaded by A. G. Woodside on November 7, 2018, at: https://medium.com/@jdelosangeles/project-framework-comparisons-agile-vs-waterfall-vs-hybrid-vs-lean-dc6801d217e4

Kao, Y.-F., Shen, Y.-Z. (2009). The effects of social entrepreneurial personality, entrepreneurial motivation and personal abilities on relationship network—a case study. *J. Entrepreneurship Res. 4*(4), 29–55.

Kumar, V., & Sharma, R. R. K. (2017). An empirical investigation of critical success factors influencing the successful TQM implementation for firms with different strategic orientation. *International Journal of Quality & Reliability Management, 34*, 1530–1550.

Lindner F and Wald A (2011), Success factors of knowledge management in temporary organizations. *International Journal of Project Management, 29*(7), 877–888.

LaPorte, T., & Consolini, P. M. (1991). Working in practice but not in theory: Theoretical challenges of "high-reliability organizations". *Journal of Public Administration Research and Theory: J-PART, 1*(1), 19–48.

Malach-Pines, A., Dvir, D., & Sadeh, A. (2009). Project manager-project (PM-P) fit and project success. *International Journal of Operations & Production Management, 29*, 268–291.

Marzagão, D. S. L., & Carvalho, M. M. (2016). Critical success factors for Six Sigma projects. *International Journal of Project Management, 34*, 1505–1518.

Obstfeld, D. (2005). Social networks, the tertius iungens orientation, and involvement in innovation. *Administrative Science Quarterly, 50*, 100–130.

Pal, R., Wang, R., & Liang, X. (2017). The critical factors in managing relationships in international engineering, procurement, and construction (IEPC) projects of Chinese organizations. *International Journal of Project Management, 35*, 1225–1237.

Powell, T. C., Lovallo, D., & Fox, C. R. (2011). Behavioral strategy. *Strategic Management Journal, 32*, 1369–1386.

Ragin, C. C. (2000). *Fuzzy-set social science*. Chicago: University of Chicago Press.

Ragin, C. C. (2008a). *Redesigning social inquiry: Fuzzy sets and beyond*. Chicago: Chicago University Press.

Ragin, C. C. (2008b). Measurement versus calibration: A set-theoretic approach. In *The Oxford handbook of political methodology* (p. 174e198). Oxford, UK: Oxford University Press.

Ramesh, B., & Tiwana, A. (1999). Supporting collaborative process knowledge management in new product development teams. *Decision Support Systems, 27*, 213–235.

Rodriquez, A. G., & Williams, T. M. (1998). System dynamics in project management: Assessing the impacts of client behavior on project performance. *Journal of Operations Research Society, 49*, 2–15.

Seashore, S E (1954). *Group cohesiveness in the industrial work group* Ann Arbor University of Michigan Press.

Simon, H.A. (1978). Rationality as process and as product of thought. American Economic Review, 68:1–16.

Song, S., Nerur, S., & Teng, J. (2007). An exploratory study on the roles of network structure and knowledge processing orientation in the work unit knowledge management. *Advances in Information Systems, 38*(2), 8–26.

Song, S., Nerur, S., & Teng, J. T. C. (2008a). An exploratory study on the roles of network structure and knowledge processing orientation in work unit knowledge management. *The DATA BASE for Advances in Information Systems, 38*, 8–26.

Song, S., Nerur, S., & Teng, J. T. C. (2008b). Understanding the influence of network positions and knowledge processing styles. *Communications of the ACM, 51*, 123–126.

Stewart, G. L., & Barrick, M. (2000). Team structure and performance: Assessing the mediating role of intrateam process and the moderating role of task type. *Academy of Management Journal, 43*, 135–148.

Teo, M. M., & Loosemore, M. (2017). Understanding community protest from a project management perspective: A relationship-based approach. *International Journal of Project Management, 35*, 1444–1458.

Tiwana, A. (2008). Do bridging ties complement strong ties? An empirical examination of alliance ambidexterity. *Strategic Management Journal, 29*, 251–272.

Tsai, W. (2001). Knowledge transfer in intra-organizational networks: Effects of network position and absorptive capacity on business unit innovation and performance. *Academy Management Journal, 44*, 996–1004.

Van der Vegt, G. S., Emans, B. J. M.,&Van de Vliert, E. 2000. Affective responses to intragroup interdependence and job complexity. *Journal of Management, 26*:633–655.

Wallace, L., Keil, M., & Rai, A. (2004). Understanding software project risk: A cluster analysis. *Information & Management, 42*(1), 115–125.

Wasserstein, R. L., & Lazar, N. A. (2016). The ASA's statement on p-values: Context, process, and purpose. *The American Statistician, 70*, 129–133.

Wasserstein, R. L., & Lazar, N. A. (2017). The ASA's statement on p-values: Context, process, and purpose. *The American Statistician, 70*, 129–133.

Weick, K. (1987). Organizational culture as a source of high reliability. *California Management Review, 29*, 112–127.

Weick, K. (2007). The generative properties of richness. *Academy of Management Journal, 50*(1), 14–19.

Weick, K. E. S., Sutcliffe, K. M., & Obstfeld, D. (2008). Organizing for high reliability: Processes of collective mindfulness. *Crisis Management, 3*(1), 81–123.

Wiig, K. M. (1997). Knowledge management: An introduction and perspective. *Journal of Knowledge Management, 1*, 6–14.

Woodside, A. G. (2013). Moving beyond multiple regression analysis to algorithms: Calling for a paradigm shift from symmetric to asymmetric thinking in data analysis and crafting theory. *Journal of Business Research, 66*, 463–472.

Woodside, A. G. (2017). Releasing the death-grip of null hypothesis statistical testing (p< .05): Applying complexity theory and somewhat precise outcome testing (SPOT). *Journal of Global Scholars of Marketing Science, 27*, 1–15.

Woodside, A. G., Nagy, G., & Megehee, C. M. (2018). Applying complexity theory: A primer for identifying and modeling firm anomalies. *Journal of Innovation & Knowledge, 3*, 9–25.

Yang, L.-R., Chen, J.-H., & Huang, C.-F. (2012). Requirements definition and management practice to improve project outcomes. *Journal of Civil Engineering and Management, 18*(1), 114–124.

Young, R., & Poon, S. (2013). Top management support—almost always necessary and sometimes sufficient for success: Findings from a fuzzy set analysis. *International Journal of Project Management, 31*, 943–957.

Zadeh, L. (1965). Fuzzy sets. *Information and Control, 8*(3), 338–353.

Zadeh, L. (1996). Fuzzy logic: Computing with words. *IEEE Transactions on Fuzzy Systems, 4*, 103–111.

Zadeh, L. (2010). Lofti Zadeh on CWW. In J. M. Mendel, L. A. Zadeh, E. Trillas, R. Yager, J. Lawry, H. Hagras, & S. Guadarrama (Eds.), *What computing with words means to me. IEEE computational intelligence magazine (February, 20–26).* https://people.eecs.berkeley.edu/~zadeh/papers/What%20Computing%20with%20Words%20Means%20to%20Me-CIM%202010.pdf. Accessed 8 Oct 2016.

Ziliak, S. T., & McCloskey, D. N. (2008). *The cult of statistical significance: How the standard error costs us jobs, justice and lives*. Ann Arbor: University of Michigan Press.

Chapter 3
Accurate Outcome Performance Screening in Strategic Management

Gábor Nagy, Carol M. Megehee, and Arch G. Woodside

Abstract Firm's operating contexts and asymmetric perspectives of success versus failure outcomes are two essential features typically absent in research on firms' implemented strategies. The study here describes and provides examples of formal case-based models (i.e., constructing algorithms) of firms' implemented strategies within several of 81 potential context (task environments) configurations—large versus small, service versus production orientation, low versus high competitive intensity, low versus high technological turbulence, and ambiguous settings for each. The study applies the tenets of complexity theory (e.g., equifinality, causal asymmetry, and single causal insufficiency). The study proposes a meso-theory and empirical testing position for solving "the crucial problem in strategic management" (Powell et al. Strategic Manage J. 32(13):1370, 2011)—firm heterogeneity—why firms adopt different strategies and structures, why heterogeneity persists, and why competitors perform differently. A workable solution is to identify/describe implemented executive capability strategies that identify firms in alternative specific task environments which are consistently accurate in predicting success (or failure) of all firms for specific implemented capabilities/context configuration. The study shows how researchers can perform "statistical sameness testing" and avoid the telling weaknesses and "corrupt practices" of symmetric tests such as multiple regression analysis (Hubbard, 2015) including null hypothesis significance testing (NHST). The study includes testing the research issues using survey responses of

G. Nagy (✉)
INSEEC Paris Business School, Department of Marketing, Paris, France
e-mail: gnagy@inseec.com

C. M. Megehee
Coastal Carolina University, Wall College of Business Administration, Conway, SC, USA
e-mail: cmegehee@coastal.edu

A. G. Woodside
Yonsei University, Yonsei Frontier Lab, Seoul, Republic of Korea
e-mail: arch.woodside@bc.edu

A. G. Woodside (ed.), *Accurate Case Outcome Modeling*,
https://doi.org/10.1007/978-3-030-26818-3_3

405 CEO and chief marketing officers in 405 Hungarian firms. The study describes algorithms indicating success cases (firms) as well as failure cases via deductive, inductive, and abductive fuzzy set logic of capabilities in contexts solutions.

Keywords Configurations · Marketing capabilities · Marketing execution capabilities · Firm performance · Subjective firm performance

1 Introduction

The study extends Herbert Simon's (1990) metaphor that human rational behavior is shaped by two blades of scissors (i.e., the structure of task environments and the computational capabilities of the actor) analogy to represent the two blades of scissors that influence firms' implemented strategy: design/execution of marketing capabilities and the context of the firm's operational environment. Not describing, understanding, or modeling the contingencies that surround the capabilities to implement marketing strategies through actions in specific marketplaces plague extant research on implemented strategy (Piercy 1998). Mainly, shortcomings stem from viewing capabilities as either stable over time or as company-wide (Collis 1994; Martin and Eisenhardt 2010). Thus, prior studies devote less attention toward analyzing marketing implementation concerning process within firms and place a higher emphasis on understanding structural alignment and administrative control (Day 1994, 2011; Vorhies and Morgan 2003; Walker and Ruekert 1987). This currently dominant research approach occurs because marketing implementation issues have been separated from the strategy process formulation (i.e., a "formulation-implementation dichotomy" exists) (Piercy 1998; Sashittal and Jassawalla 2001; Srivastava et al. 1998). To move toward a process and choice-based view of market capability implementation, researchers need to use theory and methods that can cope with and replicate organizational and environmental complexities (Collis 1994; Vorhies and Morgan 2003; Walker and Ruekert 1987).

The practical implementation of planned marketing strategy is key to linking marketing efforts with performance (Olson et al. 2005; Vorhies and Morgan 2003). Effective implementation can be tricky, as it requires coordinated efforts of individuals throughout the organization, and (senior) managers should act as conductors to define success factors contributing to strategy implementation and to develop an organizational system that promotes those activities (Olson et al. 2005). However, researchers conclude from their findings that implementing planned marketing strategy is a managerial task that consumes substantial time and resources but ends in failure frequently (Noble and Mokwa 1999; Sashittal and Jassawalla 2001; Thorpe and Morgan 2007). Such negative outcomes can be credited to failures stemming from the ways how businesses organize their marketing activities (McKee et al. 1989; Ruekert and Walker Jr 1987). Despite the theoretical and managerial importance of this issue, researchers know little about how capabilities and capability implementation should be organized to enable business strategy implementation and how it affects performance (Walker and Ruekert 1987; Workman Jr. et al. 1998).

Investigating this complex theoretical and managerial problem presents two substantial challenges. First, coordination marketing activities (capabilities and capability execution) with strategic goals are viewable as a multidimensional phenomenon consisting of many interrelated elements (Walker and Ruekert 1987). Therefore, evaluating the complex relationships of these terms requires a holistic view making up marketing capabilities, execution of marketing capabilities, and organizational contingencies (Walker and Ruekert 1987). Second, successful strategy implementation requires reconciling multiple and often conflicting contingencies (e.g., Ruekert et al. 1985), and the full range of possible contingencies makes the identification of dynamic configurations of marketing organization variables necessary to implement a business strategy seemingly intractable (Vorhies and Morgan 2003). Therefore, a major challenge for researchers is to assess whether or not marketing activities are organized in ways that enable the implementation of a strategy that achieves financial and marketing objectives successfully (Walker and Ruekert 1987).

The study here recognizes this knowledge gap and provides three contributions. First, the study illustrates how marketing capabilities align with marketing capability execution in different organizational contexts. Second, applying Simon's (1990) metaphor, the study articulates how the complex configurations of marketing capabilities, marketing capability execution, and different (internal and external) organizational contexts can affect high versus low firm profit margin, market performance, and customer performance. The study indicates the implications of the relationships among marketing capabilities, marketing capability execution, and organizational context, for high versus low business performance—in both theory and practice of strategic marketing.

Rather than constructing and testing directional-relationship hypotheses using symmetric linear modeling, the study here contributes by describing, explaining, and predicting marketing implemented strategies and outcomes by applying complexity theory tenets and asymmetric configurational modeling. The following sections include brief explanations of complexity theory tenets and asymmetric modeling steps. The steps and the findings of the study support the usefulness of recognizing that in discretizing and cross-tabulations of cases by quintiles (or terciles), for most independent variables (Xs) and a dependent variable (Y) results in cases appearing in all 25 cells (or 9 cells for terciles)—given that a study includes a reasonably large number of cases ($n \geq 100$). Consequently, researchers can improve their study's contributions by using "four-corner outcome analysis" (Nagy et al. 2018) to examine when cases with high scores for X associate with high Y scores (X•Y), when cases with low scores for X associate with high Y scores (~X•Y), when cases with high scores for X associate with low Y scores (X•~Y), and when cases with low scores for X associate with low Y scores (~X•~Y). (The mid-level dot, "•," indicates the Boolean logical "AND" operation; the sideway tilde symbol, "~," indicates negation, that is, a low score.) Rather than constructing directionality relationship models with only one dependent-variable per model, the present study also shows that such four-corner outcome analysis is useful for modeling case outcomes for alternative configurational conditions (e.g., firms experiencing low versus high sales in combination with low versus high customer satisfaction). The middle of Fig. 3.1 illustrates such four solutions for two configurational outcomes as well as eight solutions for three configurational outcomes.

Figure 3.1 is a visual of organizational (firm) context capabilities, two blades, influencing performance outcomes. Fig. 3.1 includes three general propositions.

> P1: Two-way influences occur between firm demographics and environmental turbulence. P2: Two-way influences occur between firm marketing organizing capabilities and marketing intelligence/knowledge capabilities. P3: Firm performance outcomes depend on specific configurational conditions that include firm ecology (context) and firm capabilities

The sections following introduction describe these propositions and present an empirical examination of P3.

Following this introduction this chapter has the following structure. Section two examines the configurational nature and scope of marketing capabilities and marketing execution capabilities (or implementation) in four firm contingencies (large versus small, service versus production orientation, low versus high competitive intensity, and low versus high technological turbulence) and ambiguous settings for each. Section three provides a framework that links marketing organizational capabilities to marketing execution capabilities (or implementation) and firm context (task environment) to eight possible complex outcomes. These eight include high profit and high market performance (or high customer performance), high profit and low market performance (or low customer performance), low profit and high market performance (or high customer performance), and low profit and low market performance (or low customer performance). Section four closes with a discussion of the theoretical, practical, and pedagogical contributions of the study.

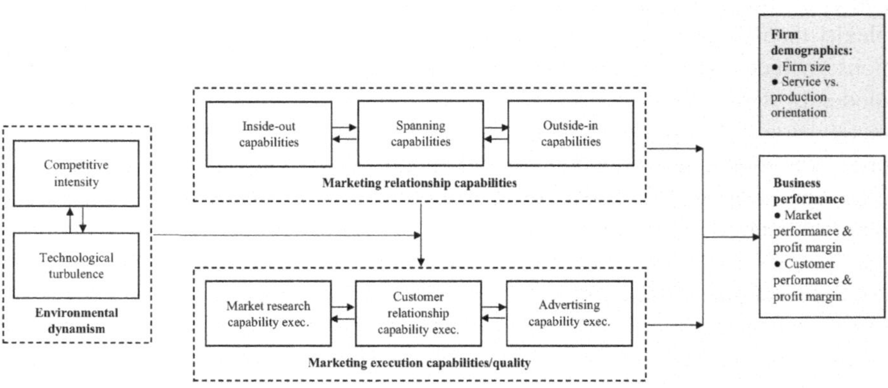

Fig. 3.1 Conceptual model

2 Configurational Nature of Marketing Capabilities and Organizational Contexts

2.1 Configuration Elements of Marketing Organizational Capabilities and Marketing Execution Capabilities Fit with Organization Context

Configuration theory and strategic marketing literature outline three crucial constructs that are relevant to understanding and assessing marketing organization fit with organizational contexts: marketing organizational capabilities, marketing execution capabilities, and organizational context. Capabilities govern the transformation of inputs into outputs (i.e., enhanced resource deployment). Effective capability executions contribute to unique product market position and competitive advantage. Capabilities represent resources accumulated over time that are not tradable on factor markets, anticipating the condition of inimitability (Vorhies et al. 1999). Day (1994) identifies three types of (marketing) capabilities underlying firm marketing processes that enable a firm to provide superior value to customers: inside-out capabilities, outside-in capabilities, and spanning capabilities. *Inside-out capabilities* are activities that are activated by market requirements, competitive challenges, and external opportunities (e.g., manufacturing and other transformation activities, logistics, and human resource management, including recruiting, training, and motivating employees) (Day 1994). *Outside-in capabilities* almost exclusively focus outside the organization (e.g., marketing sensing, customer linking, channel bounding, and technology monitoring). Their purpose is to connect other organizational capabilities to the external environment and "enable the business to compete by anticipating market requirements ahead of competitors and creating a durable relationship with customers, channel members, and suppliers" (Day 1994, p. 41). *Spanning capabilities* are needed to integrate the inside-out and outside-in capabilities (e.g., strategy development, new product/service development, price setting, purchasing, and customer order fulfillment).

Besides marketing organizational capabilities, a firm's senior and line executives need skills that enhance the implementation of the firm's processes and activities that flow from marketing strategy formulation. The core objective of a marketing organization is to attract and retain customers by providing them superior value compared to the organization's main rivals (Vorhies and Morgan 2003). For achieving this goal, the organization should design and execute practices (i.e., business or operating processes) that enhance the implementation of marketing strategy objectives. These processes require an integrated sequence of related work tasks to accomplish organizational goals. Although the literature identifies several processes that enhance marketing execution capabilities, the following three are viewable as particularly crucial in delivering high customer value. Execution of market research capabilities helps a firm to enhance its understanding of customers' current and future wants and needs with, for example, customer surveys, focus groups, concept testing, and conjoint analysis (Leonard and Rayport 1997).

Advertising capability execution is considered a marketing capability as it combines and converts tangible and intangible inputs into outputs (Ratnatunga and Ewing 2005). Customer relationship capability execution guides the organization's attitude toward initiating, maintaining, and terminating customer relationships (Javalgi et al. 2006).

The nature of the marketing activities undertaken and the ways they are performed are definable by the task characteristics of marketing organization (Day 1999). The system-structural view perspective provides a good understanding of the relationship between marketing organizations task and structural characteristics and performance (Ruekert et al. 1985; Walker and Ruekert 1987). Among the different task characteristics identified in the literature, firm size is considered the most fundamental organizational characteristic. Firm size can affect the number of resources available for gathering and processing market information (Appiah-Adu and Singh 1998), the culture of the organization (Pelham and Wilson 1996), a firm's chosen target market (Cohn and Lindberg 1974), and the aggressive steps taken by the firm. Service versus production orientation can affect the quality of the value offer (Anderson et al. 1997), the means of profit maximization (Anderson et al. 1997), and the process and resources structure of consumer needs satisfaction (Edvardsson et al. 2000). Competitive intensity is the degree of competition, including the aggressiveness and the breath of aggressive steps (Jaworski and Kohli 1993). Technological turbulence refers to the predictability of the technological changes in a given product market (Jaworski and Kohli 1993). Together these characteristics indicate the ability of the marketing organization to perform necessary marketing activities.

2.2 Marketing Organizational Capabilities and Their Fit with Organizational Context and Performance

The practical implementation of a planned marketing strategy is prerequisite to linking marketing activities with performance (Vorhies and Morgan 2003). The firm needs organizational capabilities and execution capabilities for achieving marketing strategy goals that are implementable to achieve high firm performance (Walker and Ruekert 1987; Workman Jr. et al. 1998). The strategic marketing literature suggests that different strategic goals require marketing organizations with diverse marketing organizational capabilities, marketing execution capabilities, and varying configurations of structural and task characteristics (Walker and Ruekert 1987). Therefore, marketing activities need organizing in ways that they fit with business strategy, organizational structure, and task characteristics for achieving financial and marketing performance objectives (Walker and Ruekert 1987).

When resources and related activity systems complement one another, their potential to create a sustainable competitive advantage is enhanced (Collis and Montgomery 1998; Milgrom et al. 1991). The magnitude of the firm's competitive advantage is the function of the value, rareness, inimitability, and no-substitutability of its resources and capabilities (Conner and Prahalad 1996). In support of this, firms can deliver high customer value, achieve a distinct market position, and/or neutralize the threat posed by their competitors if the resources and capabilities they exploit are narrowly available and organized in ways that they fit with related activities, organizational structure, and task characteristics (Walker and Ruekert 1987).

However, what constitutes superior marketing performance may differ along various strategic goals. A sound strategy may focus on proactive sales expansion for increasing sales volume and market share (i.e., short-term orientation) (Reichheld 1993). In contrary, a customer-focused firm emphasizes the increasing importance of customer satisfaction and loyalty, which stimulates the organization to search for ways to better serve customers and sacrifices short-term profit-seeking behavior for long-term customer relationships (i.e., long-term orientation) (Jayachandran et al. 2004). However, increasing a firm's market share along with boosting sales volume and increasing customer satisfaction and customer loyalty may conflict. Because these two objectives do not converge and can relate inversely in the short (and long) run, a firm must make crucial trade-off decisions when establishing marketing goals and assign resources (e.g., Morgan et al. 2002; Walker and Ruekert 1987). Therefore, assessing marketing organization fit with marketing strategy goals requires identification of different configurations for firms seeking to grow either sales volume and market share or customer satisfaction and customer loyalty of their marketing performance (Jayachandran et al. 2004; Walker and Ruekert 1987).

2.3 *Marketing Capabilities and Their Fit with Organizational Context and Market Performance and Profit Margin*

2.3.1 High Market Performance and High-Profit Margin

Firms that produce and sell high-quality products and services can charge higher prices and tend to have a higher market share (Buzzell et al. 1975). Moreover, consumers can perceive market share as a signal for product/brand quality and their widespread acceptance as an indicator for superior product quality that may lead to higher-profit margin (Smallwood and Conlisk 1979; Buzzell et al. 1975). The relationship between market share and profitability is grounded in several theories. Specifically, as market share increases, a business is likely to have declining marketing costs as a percentage of sales, higher economies of scale, shrinking purchases-to-sales ratio, lowering input costs, or getting extra concessions from channel members (Demsetz 1973).

2.3.2 High Market Performance and Low-Profit Margin

A significant market share may not necessarily lead to high profits, as business opportunities promising high returns entail high risks. Firms with high market share are more likely to be exposed to competitors, consumers, and the larger operational environment as these actors are more likely to take specific actions against high-share companies than against their small-share counterparts (Bloom and Kotler 1975). Buzzell et al. (1975) contend that as market share rises, turnover may rise slightly, but turnover may not be a significant factor contributing to higher rates of return. Moreover, the PIMS ("Profit Impact of Market Strategies") study shows that firms with a market share above 40 percent earn three times more than firms with market shares under 10 percent. However, there is an optimal market share in a given product/market, and a departure from this optimal share (in either direction) would alter the firm's long-run profitability in an unsatisfactory way (Bloom and Kotler 1975).

2.3.3 Low Market Performance and Low-Profit Margin

Market share provides certain advantages to firms, for example, brand recognition, economies of scale, and greater bargaining power with suppliers and channel members (Woo and Cooper 1981). However, certain conditions mitigate against the favorable impact of market share on profitability, such as demand for frequent product changes (Jaworski and Kohli 1993), rapid technological changes (Jaworski and Kohli 1993), the impact of the regulatory environment (Bloom and Kotler 1975), and the high cost of winning over customers with strong brand loyalties to competitive products (Carpenter and Nakamoto 1989; Keller 1993). In Woo and Cooper's (1981) study, ineffective low-share businesses followed competitive strategies of high-share businesses, with a broad product line, high engagement in R&D activities, a more intensive marketing posture (at higher levels of vertical integration), an insufficient scale of operation to support all activities and insufficient resources. Overall, this reflects high-quality, medium price policy complemented by careful spending on marketing, R&D, and vertical integration that eventuates in low market share and low profitability (Woo and Cooper 1981).

2.3.4 Low Market Performance and High-Profit Margin

The classical Schumpeterian view posits that an innovative new product faces low competition and earns relatively high profits, but high-profit rates attract imitators, and the level of competition faced by the product translates into reduced profits (Schumpeter 1950). However, Schumpeter does not contend that the profit profile of every firm must necessarily follow the same course. For example, firms with younger product portfolios may still be in relative monopoly positions, while older products may have exposure to much higher levels of competition (Roberts 1999; Schumpeter 1950).

2.4 *Marketing Capabilities and Their Execution Fit with Organizational Context and Customer Performance*

2.4.1 High Customer Performance and High-Profit Margin

For firms that achieve a high level of customer satisfaction, attracting new customer might be less costly (Fornell 1992), as satisfied customers are more likely to engage in positive word of mouth about the firm (Reichheld and Sasser Jr 1990), and advertising agencies might also be more willing to work with a successful firm as claims of satisfied customers make advertising more effective (Anderson et al. 1994). Furthermore, an enhanced reputation can support new product introductions by providing instant awareness and lowering the buyer's risk of a trial (Robertson and Gatignon 1986; Schmalensee 1978). A good reputation can also help a firm in establishing and maintaining relationships with key suppliers and potential allies (Anderson and Weitz 1989). Thus, customer satisfaction and customer loyalty may well contribute to high profitability.

2.4.2 High Customer Performance and Low-Profit Margin

An extreme, single-minded pursuit of customer satisfaction may lead to overspending and misallocation of resources (Rust et al. 1995). Prior research supports a positive relationship between perceived product quality and market share (see Buzzell and Wiersema 1981). However, if customers do not appreciate greater conformance to product specifications, a firm may be unable to charge higher prices (Szymanski et al. 1993). Alternatively, the usage of more expensive features to enhance product quality may lead to higher prices and reduced sales (and shrinking profits), if consumers do not accept higher prices (Szymanski et al. 1993). Moreover, increasing the net effect of product quality to enhance customer satisfaction (and loyalty) may be difficult to predict, as superior quality requires more innovative components, less standardized procedures, and increased marketing efforts. If these extra costs cannot be passed on to consumers (eventuating in high customer satisfaction), profit margins would squeeze, and quality and profit would be inversely related (Phillips et al. 1983; Szymanski et al. 1993). Thus, high customer satisfaction and loyalty do not necessarily lead to high profitability.

2.4.3 Low Customer Performance and Low-Profit Margin

Low customer satisfaction implies higher turnover of the customer base, higher replacement costs, and higher customer acquisitions costs (Anderson et al. 1994). Customers with low satisfaction buy less frequently, purchase in smaller values, and may not be prone to try other offerings by the firms, thus not increasing revenues and not contributing to lowering the costs of marketing and sales (Reichheld and Sasser Jr 1990). However, it may be challenging for firms to "get rid of" such customers. There are several reasons for this. (1) There are only a few viable product/brand alternatives in the market and customers do not have many options to choose from (Lee et al. 2001). (2) Due to high switching costs, customers may find it difficult (or do not want) to change to other firms (Fornell 1992; Reinartz and Kumar 2002). (3) Dissatisfied customers can undermine the firm's overall reputation and spread negative word of mouth about the firm that increases the cost of attracting new customers (Fornell 1992). These factors may increase costs associated with handling dissatisfied customers that can finally lower profitability.

2.4.4 Low Customer Performance and High-Profit Margin

Relatively low customer satisfaction and high economic returns are associated with firms competing in markets (or industries) with limited competition (Anderson et al. 1997). For these firms, satisfied customers provide fewer benefits. In this highly concentrated industry (or market), one firm (or a small set of firms) has a secure location advantage, giving a firm a degree of monopoly power that allows a firm to earn high profits (Anderson et al. 1997). Furthermore, firms that do not invest in satisfaction enhancement activities (because they face limited competition) might have more unused resources available (i.e., many other high net present value projects available), and these cost savings can be invested in other processes/activities that may have a positive impact on profitability (Mittal et al. 2005). Thus, low customer satisfaction and high profit can co-occur.

2.5 The Effect of a Competitive Environment

Reputation (stemming from customer satisfaction and loyalty) can provide a halo effect for the firm positively influencing customer evaluations by providing shelter (or isolation) from short-term shocks (i.e., aggressive moves) in the environment (Anderson et al. 1994). However, in competitive markets firm can attract customers with a relatively low level of customer satisfaction from their rivals (Anderson and Mittal 2000). Also, growing technological turbulence induces more opportunities to create value for buyers (Narver and Slater 1990). Thus, product orientation may require some degree of monopoly protection (or a market with limited competition), where—due to switching barriers—firms can get away with a lower level of customer satisfaction (Fornell 1992). In contrary, when switching barriers are becoming less critical, it becomes more imperative to retain current customers through customer satisfaction (Fornell 1992).

2.6 The Effect of Firm Demographics

2.6.1 Firm Size

The literature on small firms suggests certain (market share) strategies. Generally, smaller firms are advised not to meet larger competitors head-on (Woo and Cooper 1981). Smaller firms also concentrate on specialized products, localize business operations, and pursue high-quality versus high-volume strategy (Appiah-Adu and Singh 1998). Furthermore, smaller firms also benefit from the provision of customer services, (continuous) product innovation, and a narrow product line (Cohn and Lindberg 1974). In contrary, larger firms have enough resources to face more significant competitors face-to-face (Hall 1980), market a broader product portfolio (Prajogo et al. 2013), and may compete in several markets (simultaneously) while selling in high volumes (Cohn and Lindberg 1974).

2.6.2 Service Versus Production Orientation

Products and services are fundamentally different in how they are produced and consumed (Bateson and Hoffman 1999; Grönroos 1990). Firms and organizations typically produce services that are in an interactive process in which customers play an active role, while firms manufacture products typically in advance of their purchase and consumption (Edvardsson et al. 2000). A service company offers the prerequisites for a service (i.e., the means of realization in the form of process and resources structure) to guarantee service quality, and consistent quality needs to be present in all parts of the prerequisites for the service (Edvardsson et al. 2000). In contrast, products are the results of a sorting process that includes the separation and accumulation of parts and materials. The combinations of these parts and materials meet the needs of production units and ultimately get into production (Johnson et al. 1999). Thus, unlike production-oriented firms, service firms deliver high business performance if the integrity across processes and activities is high.

3 Data and Methods

3.1 Data

This study uses a sample of 405 Hungarian firms. The sampling frame was based on US Standard Industrial Classification (SIC) codes, following US industry-specific and area-specific classifications. The sampling frame was obtained from Dun & Bradstreet's company directory of Hungarian firms, which contained a pool of 18,293 B2B and B2C firms (B2B and B2C firms were equally represented in the sampling frame). The data were collected using survey sent to the managers responsible for marketing operations. 5000 questionnaires were sent to chief marketing executives to a sample with equal distribution on B2B and B2C firms (50–50%). A second wave of requests to nonrespondents followed. Nonresponse bias was tested comparing early versus late respondents (Armstrong and Overton 1977)—no statistically significant differences occurred in construct measures. After data cleaning, 405 usable questionnaires were available (B2B = 145, B2C = 260) that correspond to a response rate of 8.1%. Per firm size and industry classification, the final sample stratifies into the following set of firms: (1) ≤49 = 171, 50–99 = 105, 100–299 = 81, 300 ≥= 48); (2) consumer products = 93, consumer services = 74, business products = 68, and business services = 74, and others = 93) (for more details about sample distribution, see Appendix A). Considering the relative complexity of the measurement instrument (eight pages) and the fact that the respondents were top executives, the response rate achieved is acceptable.

To assess discriminant validity of the latent variables, confirmatory factor analysis with AMOS 25.0 was employed. A 29-item measurement model with 11 latent constructs was built. Two items were removed from the initial factor structure due to low factor loadings (<0.65) or high modification indices. The purified model with the remaining 26 items indicated a good fit to the data with χ^2/df = 1.539 ($p < 0.001$), CFI = 0.969, NFI = 0.917, GFI = 0.932, and RMSEA = 0.037. All items' loadings on their corresponding latent construct were significant at the 0.05 level, demonstrating adequate convergent validity. By assessing discriminant validity, the procedure outlined byFornell and Larcker (1981) was applied. An examination of AVE showed that for each pair of variables, the extracted variance was greater than the squared correlations between the constructs. As such, the study achieves high discriminant validity. Internal reliability of scales was assessed with Cronbach's alpha and each construct (except for customer relationship capability execution and technological turbulence, Cronbach's alpha = 0.67 and 0.63, respectively) exceeded the threshold value of 0.80. Based on CFA results latent constructs were created by summing item scores in the respective measurement instrument to retain high variance within the instruments and eliminate shared variance due to possible overlap across constructs (Hair et al. 2006). Common method variance (CMV) was tested following Harman's one-factor test. A non-rotated principal component analysis identified ten factors with eigenvalues greater than one, which together explains 65.73% of the total variance, with no single factor accounting for more than 50% of the variance. Additionally, the study employs the marker variable test to control for common method variance (Podsakoff et al. 2003). To make sure common method variance is not an issue with the sample data, a four-item measure was used to measure "internal marketing support assets"[1]. The correlation between the marker variable and the theoretically independent variable is 0.18, further suggesting that common method variance is less of a threat to the validity of results (Lindell and Whitney 2001). Appendix B includes the items for measuring the constructs in the study.

[1]The items for this five-point Likert scale include cost-effective production, advanced marketing information system, advanced cost-controlling system, and patents and licenses.

3.2 Analysis

The current study employs a set-theoretic approach based on fuzzy set QCA. The set-theoretic approach allows for a detailed analysis of how causal conditions contribute to an outcome of interest. QCA treats cases as configurations of attributes (resembling overall types), and the comparison of cases can allow a researcher to omit attributes that are unrelated to the outcome of interest. QCA analysis examines causal patterns by focusing on the set-subset relationship. For a specific outcome (e.g., high performance), it examines members of the set of "high-performing" organizations and then identifies that combinations of attributes are associated with the outcome of interest using Boolean algebra and algorithms (that allow logical reduction of numerous, complex causal conditions into a reduced set of configurations that lead to the outcome) (Fiss 2011). Although more widespread in management research, methods as moderated regression analysis, clustering algorithms, latent class analysis, and the deviation score approach all have their notable limitations as they assume linearity, additive effects, and unifinality (Fiss 2007; Woodside 2013).

The causal complexity in configuration theory as well as the study's set-theoretic assumptions—necessary and sufficient conditions, equifinality, and causal ambiguity—call for using QCA (Fiss 2007). A set-theoretic approach allows to study (1) equifinality (i.e., multiple paths lead to the same outcome), (2) nonlinearity (i.e., "relationships between variables can be nonlinear with abrupt switches occurring, so the same >> cause << can, in specific circumstances, produce different effects") (Urry 2005, p. 4), and (3) the tipping-point tenet (e.g., in strategic management models replacing a negative with a positive response to one issue in a string (path) of questions within a given complex configuration of antecedent conditions) (Urry 2005). This study builds on configuration theory, according to which, for sets of causal factors, combinations of factors were expected to contribute to effectiveness maximization.

The lowest acceptable consistency level for solutions was set equal to 0.80, when possible (the minimum threshold level recommended by Ragin (2008) is 0.75). Also, the minimum acceptable solution frequency was set at 2.0. The current study uses the truth table algorithm which is based on counterfactual analysis of causal conditions allowing for categorization of causal conditions into the core and peripheral causes. For this, a counterfactual analysis was performed that distinguishes—based on parsimonious and intermediate solutions—"easy" and "difficult" counterfactuals (Fiss 2011; Ragin 2008). Moreover, the tenet of core and peripheral causes was used to enrich the studying of equifinal configurations through the notion of neutral permutations of a given configuration (cf. Gresov and Drazin 1997; Fiss 2011). Following Fiss (2011), within a given configuration, more than one constellation of different peripheral causes may surround the causal core condition, and the permutation of peripheral causes do not affect the overall performance of the configuration. Neutral permutations distinguish between first- and second-order equifinality, providing a fine-grained understanding of the different cause-effect relationships related to an outcome. First-order equifinality denotes equifinal configurations that exhibit different core characteristics (e.g., type A vs. type B), whereas second-order equifinality refers to neutral permutations within a given first-order equifinal type (e.g., type A_1 vs. A_2 … A_n) (Fiss 2011).

3.3 Calibration of Set Memberships

3.3.1 Outcomes

High-Profit Margin, High Market Performance, High Customer Performance In this study, one objective performance measure (profit margin) and two subjective performance measures (market performance, customer performance) were chosen as primary outcomes of interest.[2] Profit margin was calculated as pre-tax profits divided by sales. This measure was calibrated to the overall performance of the Hungarian firms in the selected sectors rather than by using a sample-dependent anchor. Profit margin was averaged across different sectors (cf. Fiss 2011). The average profit margin for the Hungarian firms in selected sectors was 4.7 percent (the median profit margin of the firms in the sample was 21.0 percent for the same period). Membership in the set of firms having low objective performance was coded 0 if a firm's profit margin was below 5.0 percent (i.e., the ≈20th percentile or lower) and was coded as 1.0 if the firm's profit margin was equal or above 50.0 percent (i.e., the ≈70th percentile or higher). As the crossover point, 25.0 percent was chosen. The median sales of the firms in the sample were 400 million HUF). Firms were fully out of the set of high performers when their turnover was equal or below 150 million HUF (i.e., ≤20th percentile) and were fully in the set when their turnover was equal or above 1300 million HUF (i.e., ≥80th percentile). The hallway mark between these thresholds was 298 million HUF. Appendix C provides details on the breakpoints for fully in membership (0.95), point of maximum ambiguity (crossover = 0.50), and fully out nonmembership for each of the conditions.

Market performance was measured with sales volume and market share. Membership in the set of firms with high market performance was coded 0 if a firm showed average or below average sales volume and market share (≤5.00, i.e., ≤20th percentile) and was coded 1 if the firm scored high on sales volume and market

[2]Because chief marketing executives self-reported objective and subjective performance data, a common method was possible. Harman's single factor test was applied to assess common method bias. In case of a severe common method bias problem, a single latent factor should account for a large portion of the variance in the sample. The non-rotated principal component factor solution suggests that common method bias is not an issue here, as the analysis of the (full) sample data and all measures results in a ten-factor solution with eigenvalues larger than 1. Furthermore, the most significant factor accounted for 19 percent of the variance, and the most significant three factors together account for 36.34 percent of the variance. Although the results do not certainly rule out the existence of common method bias, they suggest that common method bias is unlikely to affect the results in any substantive way.

share (≥8.00, i.e., ≥80th percentile). The crossover point was 5.98. Customer performance was assessed with customer satisfaction and customer loyalty. Firms were fully out of the set of high market performers when the combined effect of customer satisfaction and customer loyalty was equal or below 6.00 (i.e., ≤20th percentile) and were fully in the set when the combined effect was equal or above 8.00 (i.e., ≥80th percentile). The hallway mark between these thresholds was 6.98.

Not-High-Profit Margin, Not-High Market Performance, Not-High Customer Performance This study also sought to understand whether and how the marketing organizational capabilities and marketing execution capabilities quality (in harmony with internal and external environmental contingencies) combine in the absence of high-profit margin, high market performance, and high customer performance (i.e., membership in the set of firms with not-high business performance). This was captured by the negation of the high-profit margin, high market performance, and high customer performance fuzzy sets described above (so here, for not-high-profit margin, fully in = 7.00, fully out = 57.80, crossover = 8.98; for not-high market performance, fully in = 5.00, fully out = 8.00, crossover = 5.98; and for not-high customer performance, fully in = 6.00, fully out = 8.00, crossover = 6.98).

Performance Outcomes with Contrarian Cases To assess how marketing organizational capabilities and marketing execution capabilities quality combine in the presence and/or absence of high-profit margin, high business performance, and high customer performance, consistent with Nagy et al. (2017), complex outcomes of different high and not-high business performance outcomes were created with the "fuzzy and" function in the fs/QCA 3.0 software.

3.3.2 Marketing Organizational Capabilities and Marketing Execution Capabilities Quality

Inside-out capabilities were assessed with Day's (1994) conceptual initiative. Membership in the set of firms with low inside-out capabilities was coded 0 if a combined effect of a firm's activities triggered off by market requirements, competitive challenges, and external opportunities was equal to or below 8.00 (i.e., 20th percent or lower) and was coded as 1 if the firm scored on these activities equal or above 12.00 (i.e., 80th percentile or higher). As the crossover point, 9.98 was chosen. Spanning capabilities were measured with Day's (1994) and Theoharakis et al.'s (2009) scale. Firms with spanning capabilities less than or equal to 5.00 (i.e., ≤20th percentile) were coded as full nonmembership, and firms with spanning capabilities more than or equal to 7.00 (i.e., ≥80th percentile) were coded as full membership in the bundle of activities, such as new product/service development, customer order fulfillment, and strategy development. 5.98 was set as the crossover point. Outside-in capabilities were assessed with Day's (1994) and Theoharakis et al.'s (2009) construct. Firms were fully out of the set of high outside-in capabilities when the combined effect of customer linking, channel bonding, and relationship keeping was equal or lower to 9.00 (i.e., ≥20th percent or lower), whereas firms were fully in the set when the combined effect was equal or above 12.00 (i.e., ≥80th percentile or higher). The hallway mark between these thresholds was 11.98.

New measures were developed for market capability execution, combining insights from the literature and interviews with senior marketing personnel and marketing academics. For this study, three marketing execution capabilities (types) were chosen that align well with marketing organizational capabilities introduced above (cf. Day 1994, 2011; Theoharakis et al. 2009): (1) market research capability execution that helps firms to understand consumer needs and wants and disseminate the information within the organization so it can be used by all organizational members (Glazer 1991), (2) advertising capability execution that increases brand awareness and purchase intention (Carpenter and Nakamoto 1989; Keller 1993), and (3) customer relationship capability execution that—through transaction- or relationship-specific interactions—strengthens ties with customers, which increases satisfaction, leading to high customer loyalty and retention (Javalgi et al. 2006). The threshold for membership in the set of firms with low marketing execution capabilities was set at the 20th percentile, while the threshold for membership in the set of firms with high marketing execution capabilities was set at the 80th percentile for market research capability execution, advertising capability execution, and customer relationship capability execution. As the crossover points, the following anchors were used: 5.98, 7.98, and 7.98, respectively. These marketing execution capabilities were assessed how well firms performed these activities from an internal point of view (cf. Piercy 1998; Sashittal and Jassawalla 2001; Olson et al. 2005).

3.3.3 Firm Contingencies

In calibrating firm size, the number of employees was used following the European Commission's definition and the 2004 XXXIV Hungarian law on firm size. The result is a four-value fuzzy score: under 50 employees (0), 50–99 employees (0.33), 100–299 employees (0.67), and 300 or more employees (1). Full membership in the set of firms with high service orientation was coded as 0.99 and full nonmembership in this set as 0.01. The two dimensions for environmental dynamism (i.e., technological turbulence and competitive intensity) were assessed using Jaworski and Kohli's (1993) metric. Firms that perceive low technological turbulence were coded as full nonmembership (≤5.00, i.e., ≤20th percentile), and firms that perceive high technological turbulence were coded as full membership (≥8.00, i.e., ≥80th percentile). A crossover point of 6.98 was chosen. A "full out" threshold was set to firms with perceived competitive intensity equal or lower to 7.00 (i.e., ≤20th percentile) and "full in" threshold for firms with perceived competitive intensity equal or higher to 10.00 (i.e., ≥80th percentile). The crossover point was set to 8.98.

4 Findings

4.1 Configurations for Achieving High Market Performance and High-Profit Margin Versus Configurations for Low Market Performance and Low-Profit Margin

Table 3.2 shows the results of the fuzzy set analysis of high market performance and high-profit margin versus low market performance and low-profit margin. (Table 3.1 presents correlations for all measures used in the analysis.) Following Ragin's (2008) notion, solution tables mark high on a condition with black circles ("•") and low on a condition with the tilde sign ("•"). Furthermore, cells painted in soft gray color refer to core conditions. Their core conditions group solutions. Table 3.2 shows that the fuzzy set analysis results in three solutions for the high market performance and high-profit margin outcome (with solution consistency ≥0.79) and five solutions for the low market performance and low-profit margin outcome (with solution consistency ≥0.98). With the first outcome, one first-order (or cross-type) equifinal solution and two second-order (or within-type) equifinal solutions were obtained (solutions 1 and 2a, 2b). For low market performance and low-profit margin, five effective solutions were found, all depicting first-order equifinal recipes.

Core conditions include solution 1 indicating that firms high on inside-out capabilities, spanning capabilities, market research capability execution, customer relationship capability execution, and low on outside-in capabilities and advertising capability execution achieve high market performance and high-profit margin. This configuration pertains to large firms operating in stable environments. Solutions 2a and 2b show configurations with core conditions of high market research capability execution, large size, and service orientation. Firms in these configurations operate in environments with the high pace of technological change. Furthermore, solutions 2a and 2b depict second-order equifinality (i.e., neutral permutations within a given first-order equifinal type) as core conditions are surrounded by different peripheral causes (notably low on outside-in capabilities and low on competitive intensity for 2a), but the permutation of peripheral causes does not affect the overall performance. Also, comparing solutions 2a and 2b indicates that outside-in capabilities and competitive intensity are treatable as complements. Overall, the solution set for high market performance and high-profit margin depicts configurations where marketing organizational capabilities are deployed extensively, and marketing organizational capabilities well support activities about these execution capabilities—all these configurations are relevant in a large-firm setting.

The solution set for low market performance and low-profit margin depicts five first-order equifinal solutions. Solutions 1 and 2 indicate firms that are high on inside-out and spanning capabilities. In solution 1, firms are high on outside-in capabilities. In solution 2, core conditions for the low market performance

Table 3.1 Bivariate correlations

		1	2	3	4	5	6	7	8	9	10	11	12	13
1.	Inside-out capabilities	1												
2.	Spanning capabilities	0.16	1											
3.	Outside-in capabilities	0.28	0.13	1										
4.	Advertising capability execution	−0.04	−0.04	0.08	1									
5.	Market research capability execution	0.03	−0.04	0.07	0.42	1								
6.	Customer relationship capability execution	0.07	0.00	0.04	0.15	0.31	1							
7.	Competitive intensity	0.06	−0.03	−0.01	0.08	0.07	0.05	1						
8.	Technological turbulence	0.06	−0.02	0.00	−0.02	−0.08	0.01	0.26	1					
9.	Market performance	0.28	0.02	0.22	−0.01	0.01	0.01	−0.04	−0.06	1				
10.	Customer performance	0.23	0.02	0.30	0.07	0.11	−0.02	0.05	−0.05	0.37	1			
11.	Profit margin	0.00	−0.08	0.08	0.00	0.03	0.04	−0.06	−0.03	0.00	−0.04	1		
12.	Large firm	0.05	0.01	0.05	0.03	0.04	0.06	0.02	0.05	0.03	0.03	−0.01	1	
13.	Service orientation	−0.01	0.00	−0.09	0.01	−0.01	0.03	0.03	0.06	−0.02	0.00	0.02	−0.14	1

Note: Correlations of 0.14 or higher are significant at ≤ 0.05

Table 3.2 Configurations for achieving high market performance and high-profit margin versus configurations for achieving low market performance and low-profit margin

	High market performance & high-profit margin			Low market performance & low-profit margin				
	Solution[a, b, d]			Solution[a, c, d]				
Configuration	**1**	**2a**	**2b**	**1**	**2**	**3**	**4**	**5**
Marketing organizational capabilities quality								
Inside-out capabilities	•	•	•	•	•	~	~	~
Spanning capabilities	•	•	•	•	•	•	•	•
Outside-in capabilities	~	~	•	•	~	~	•	~
Marketing execution capabilities quality								
Advertising capability execution	~	•	•	~	~	•	•	•
Market research capability execution	•	•	•	~	~	•	•	•
Customer relationship capability execution	•	•	•	~	~	~	~	•
Environmental dynamism								
Competitive intensity	~	~	•	~	•	~	~	~
Technological turbulence	~	•	•	~	•	•	•	•
Firm demographics								
Large firm	•	•	•	~	•	~	•	•
Service orientation	~	•	•	~	~	•	~	~
Raw coverage	0.05	0.05	0.06	0.03	0.02	0.01	0.03	0.03
Unique coverage	0.04	0.01	0.02	0.01	0.00	0.01	0.01	0.01
Consistency	0.80	0.80	0.85	0.98	1.00	1.00	1.00	1.00
Overall solution coverage	**0.12**			**0.09**				
Overall solution consistency	**0.79**			**0.98**				

Note: [a]'•' means high on a condition and '~' denotes low on a condition
[b]Frequency cutoff = 2, consistency cutoff = 0.80
[c]Frequency cutoff = 3, consistency cutoff = 0.97
[d]Cells with shade denote core conditions; cells with white color indicate peripheral conditions. Causal coreness defined concerning the strength of the evidence relative to the outcome, not concerning the connectedness to other configurational elements (Fiss 2011)

and low-profit margin (outcome) are low on outside-in capabilities, low customer relationship capability execution, and production orientation. This result happens in a somewhat turbulent market environment, which indicates that the lack of customer linking, channel bonding, and relationship keeping capabilities and customer relationship capability execution most likely leads to low market performance and low-profit margin. Also, solutions 1 and 2 indicate that having robust marketing organizational capabilities but not possessing marketing execution capabilities is a recipe for consistently low market performance and low-profit margin. Solutions 3, 4, and 5 combine low inside-out capabilities, low competitive intensity, high technological turbulence, and production orientation (in the latter two solutions) for consistently low market performance and low-profit margin. This pattern may signal that even if firms possess spanning capabilities, outside-in capabilities, and capabilities that support the execution of different firm activities, in the absence of inside-out capabilities, firms will perform poorly compared to their competitors.

4.2 Configurations for Achieving High Market Performance and Low-Profit Margin Versus Configurations for Low Market Performance and High-Profit Margin

Table 3.3 shows results for a fuzzy set analysis of high market performance and low-profit margin and low market performance and high-profit margin (four first-order and two second-order equifinal solutions, solution consistency ≥0.95). Core conditions across the six solutions indicate that firms widely mobilize their capabilities and activities that support the implementation/execution of marketing activities as the external environments encourage them to do so (either competitive intensity or technological turbulence is high across the solutions, and in solutions 3a, 3b, and 5, both environmental factors take on a high value). Also, all solutions are high on service orientation. These configurations signal intensive exploitation of firm resources to increase sales volume and market share but at the same time burns profits, a condition that is unsustainable in the long term. Furthermore, 3a and 3b—two second-order equifinal solutions—show that outside-in capabilities (i.e., customer linking, channel bounding, and relationship keeping) and customer relationship capability execution can be complements. Overall, intensive exploitation of firm resources may boost sales and increases market share. However, this resource allocation significantly contributes to reduced profitability. The truth table algorithm found only one consistent solution for low market performance and high-profit margin (with solution consistency ≥0.80). Firms in this set possess (as core conditions) high spanning capabilities (i.e., new product/service development, new product/service introduction), are healthy in advertising, perform extensive market research (and effectively utilize these insights), are service oriented, have large size, operate in stable environments, and are low on customer relationship capabilities and inside-out and outside-in capabilities.

Table 3.3 Configurations for achieving high market performance and low-profit margin versus configurations for achieving low market performance and high-profit margin

	High market performance & low-profit margin						Low market performance & high-profit margin
	Solution[a, b, d]						Solution[a, c, d]
Configuration	**1**	**2**	**3a**	**3b**	**4**	**5**	**1**
Marketing organizational capabilities							
Inside-out capabilities	•	•	•	•	•	•	~
Spanning capabilities	•	•	•	•	•	•	•
Outside-in capabilities	•	•	•	~	~	•	~
Marketing execution capabilities quality							
Advertising capability execution	~	•	•	•	•	•	•
Market research capability execution	•	•	•	•	•	~	•
Customer relationship capability execution	•	•	~	•	•	•	~
Environmental dynamism							
Competitive intensity	•	~	•	•	•	•	~
Technological turbulence	~	•	•	•	~	•	~
Firm demographics							
Large firm	~	~	~	~	•	•	•
Service orientation	•	•	•	•	•	•	•
Raw coverage	0.02	0.03	0.03	0.02	0.02	0.03	0.05
Unique coverage	0.00	0.00	0.00	0.00	0.00	0.01	0.05
	0.97	0.93	0.93	0.95	0.96	0.93	0.80
Overall solution coverage	**0.08**						**0.05**
Overall solution consistency	**0.95**						**0.80**

Note: [a]'•' means high on a condition and '~' denotes low on a condition
[b]Frequency cutoff = 2, consistency cutoff = 0.80
[c]Frequency cutoff = 3, consistency cutoff = 0.97
[d]Cells with shade denote core conditions; cells with white color indicate peripheral conditions. Causal coreness defined concerning the strength of the evidence relative to the outcome, not concerning the connectedness to other configurational elements (Fiss 2011)

4.3 Configurations for Achieving High Customer Performance and High-Profit Margin Versus Configurations for Low Customer Performance and Low-Profit Margin

Table 3.4 presents the results for the fuzzy set analysis of high customer performance and high-profit margin with one first-order and three second-order solutions (solution consistency $\geq$ 0.79). Core conditions for solution 1 indicate that robust spanning capabilities, strong advertising capability execution, and strong market research capability execution coupled with the significant size and service orientation lead to high customer performance and high-profit margin. Furthermore, firms in solution 1 are low on other significant marketing organizational capabilities, don't maintain close ties with key customers, and are operating in stable environments.

Core conditions for the three second-order equifinal solutions (2a, 2b, and 2c) show recipes with robust inside-out capabilities, strong market research capability execution, strong customer relationship capability execution, and the presence of competitive market forces. Again, solutions 2a and 2b show that outside-in capabilities (i.e., customer linking, channel bonding, and relationship keeping) and advertising capability execution can be complements. In sum, solution 1 depicts (in contrary to solutions 2a, 2b, and 2c) firms can effectively satisfy their customers and make them loyal even without explicit emphasis on customer relationship management when proactive market research substitutes for other activities, innovative product design, strong service orientation, and large firm size. Compared to solutions for high market performance and high-profit margin, firms achieving high customer performance and high-profit margin seem to have core competencies in inside-out marketing processes, market research, and customer relationship management and operate in highly competitive environments.

Table 3.4 also shows the results of a fuzzy set analysis for low customer performance and low-profit margin with three first-order and two second-order solutions (solution consistency $\geq$0.99). Core condition for solutions 1 and 2a and 2b shows that firms high on marketing organizational capabilities (but low on marketing execution capabilities) perform systematically low on customer performance and profit margin. Moreover, solutions 3 and 4 depict firms with a wide range of capabilities and capability execution, but an ineffectual resource allocation. Here again—like in the case of low market performance and low-profit margin—solutions show that regardless of the quantity and configuration of necessary marketing organizational capabilities, firms lacking activities that support marketing execution capabilities are likely to experience low customer performance and low-profit margin. Also, in the absence of inside-out capabilities—although possessing a large variety of marketing organizational capabilities and processes enhancing marketing execution capabilities—firms perform poorly compared to leading competitors.

Table 3.4 Configurations for achieving high customer performance and high-profit margin versus configurations for achieving low customer performance and low-profit margin

	High customer performance & high-profit margin				Low customer performance & low-profit margin				
	Solution[a, b, d]				Solution[a, c, d]				
Configuration	**1**	**2a**	**2b**	**2c**	**1**	**2a**	**2b**	**3**	**4**
Marketing organizational capabilities quality									
Inside-out capabilities	~	•	•	•	•	•	•	~	~
Spanning capabilities	•	•	•	•	•	•	•	•	•
Outside-in capabilities	~	~	•	•	•	~	~	•	~
Marketing execution capabilities quality									
Advertising capability execution	•	•	~	•	~	~	~	•	•
Market research capability execution	•	•	•	•	~	~	~	•	•
Customer relationship capability execution	~	•	•	•	~	~	~	•	•
Environmental dynamism									
Competitive intensity	~	•	•	•	~	•	~	~	~
Technological turbulence	~	•	•	•	~	•	•	~	•
Firm demographics									
Large firm	•	~	•	•	~	•	•	•	•
Service orientation	•	~	~	•	~	~	•	~	~
Raw coverage	0.05	0.05	0.10	0.07	0.02	0.02	0.02	0.03	0.02
Unique coverage	0.02	0.01	0.05	0.03	0.01	0.01	0.01	0.02	0.00
Consistency	0.83	0.83	0.83	0.84	0.99	1.00	1.00	1.00	1.00
Overall solution coverage	**0.19**				**0.10**				
Overall solution consistency	**0.79**				**0.99**				

Note: [a]'•' means high on a condition and '~' denotes low on a condition
[b]Frequency cutoff = 2, consistency cutoff = 0.80
[c]Frequency cutoff = 3, consistency cutoff = 0.97
[d]Cells with denote core conditions; cells with white color indicate peripheral conditions. Causal coreness defined concerning the strength of the evidence relative to the outcome, not concerning the connectedness to other configurational elements (Fiss 2011)

4.4 Configurations for Achieving High Customer Performance and Low-Profit Margin Versus Configurations for Low Customer Performance and High-Profit Margin

Table 3.5 shows the fuzzy set results for high customer performance and low-profit margin. The solution set depicts four first-order (solutions 2, 4, 5 and 6) equifinal and four second-order (solutions 1a, 1b, 3a, and 3b) equifinal solutions (solution consistency ≥0.90). Solutions 1a and 1b combine high inside-out capabilities, high advertising capability execution, low customer relationship capability execution, and low firm size as core conditions. Thus, firms can get away with loose customer relationships and omit the systematic analysis of customer satisfaction and still attain high customer satisfaction and loyalty (but low-profit margin) if having strong advertising capabilities and inside-out capabilities.

The next second-order equifinal solution set (i.e., solutions 3a and 3b) shows that core conditions for high customer performance and low-profit margin are robust outside-in capabilities (i.e., customer linking, channel bounding, and customer linking), weak advertising and market research capability execution, strong customer relationship execution, and reliable service orientation. The remaining first-order equifinal solutions (i.e., solutions 2, 4, 5, and 6) depict a high level of customer relationship capability execution and service orientation as core conditions. These conditions indicate that building strong relationships through tailor-made services may have a positive effect on customer satisfaction and customer loyalty, but significantly reducing profitability (regardless of whether the firm is large/small, competes in a highly competitive/stable environment, and/or possesses a wide range of necessary marketing organizational capabilities and activities supporting execution marketing capabilities).

Table 3.5 also shows the results of a fuzzy set analysis for low customer performance and high-profit margin with three first-order equifinal solutions (solution consistency ≥0.78). Compared to solutions with high customer performance and low-profit margin, core conditions combine into sets with robust inside-out capabilities, weak outside-in capabilities, strong market research capability execution, strong customer relationship execution, large firm size, and weak/strong advertising capability execution and production/service orientation. Thus, to reach great profitability—in addition to having strong customer relationship building capabilities—firms need additional (efficiently configured) core capabilities and core activities supporting capability execution. However, these configurations lead to high-profit margin, but may not help firms to excel on customer satisfaction and customer loyalty, which can again undermine profitability in the long run.

Table 3.5 Configurations for achieving high customer performance and low-profit margin versus configurations for achieving low customer performance and high-profit margin

	High customer performance & low-profit margin								Low customer performance & high-profit margin		
	Solution [a, b, d]								Solution [a, c, d]		
Configuration	**1a**	**1b**	**2**	**3a**	**3b**	**4**	**5**	**6**	**1**	**2**	**3**
Marketing organizational capabilities quality											
Inside-out capabilities	•	•	~	~	•	•	•	•	•	•	•
Spanning capabilities	•	•	•	•	•	•	•	•	•	~	•
Outside-in capabilities	•	•	~	•	•	•	~	•	~	~	~
Marketing execution capabilities quality											
Advertising capability execution	•	•	~	~	~	~	•	•	~	~	•
Market research capability execution	•	•	•	~	~	•	•	~	•	•	•
Customer relationship capability execution	~	~	•	•	•	•	•	•	•	•	•
Environmental dynamism											
Competitive intensity	~	•	~	~	•	•	•	•	~	•	~
Technological turbulence	~	•	~	•	•	~	~	•	~	~	•
Firm demographics											
Large firm	~	~	•	•	~	~	•	•	•	•	•
Service orientation	~	•	•	•	•	•	•	•	~	~	•
Raw coverage	0.03	0.03	0.02	0.04	0.03	0.02	0.02	0.03	0.07	0.06	0.06
Unique coverage	0.03	0.01	0.00	0.10	0.01	0.00	0.00	0.00	0.02	0.01	0.04
Consistency	0.91	0.91	0.90	0.94	0.90	0.95	0.94	0.91	0.86	0.80	0.80
Overall solution coverage	**0.13**								**0.13**		
Overall solution consistency	**0.90**								**0.78**		

Note: [a]'•' means high on a condition and '~' denotes low on a condition
[b]Frequency cutoff = 2, consistency cutoff = 0.80
[c]Frequency cutoff = 3, consistency cutoff = 0.97
[d]Cells with [shaded] denote core conditions; cells with white color indicate peripheral conditions. Causal coreness defined concerning the strength of the evidence relative to the outcome, not concerning the connectedness to other configurational elements (Fiss 2011)

5 Discussion

Prior research placed high emphasis on understanding structural alignment of marketing organizational capabilities, marketing resources, and structural contingencies (Day 1994, 2011; Vorhies and Morgan 2003; Walker and Ruekert 1987) and devoted scant attention toward analyzing marketing implementation in terms of process within firms (Piercy 1998; Thorpe and Morgan 2007). The structural alignment of marketing organizational capabilities and marketing execution capabilities requires reconciling multiple and often conflicting contingencies (Ruekert et al. 1985; Vorhies and Morgan 2003; Walker and Ruekert 1987), and scholars call for investigating how marketing organizational capabilities and marketing execution capabilities/implementation can lead to high business performance using methods that can adequately handle organizational complexity (Collis 1994; Vorhies and Morgan 2003; Walker and Ruekert 1987). Despite these attempts, knowledge is still limited about how marketing organizational capabilities and marketing execution capabilities work together in different organizational settings producing different (often conflicting) performance outcomes. This study integrates prior literature on marketing organizational capabilities and marketing execution capabilities and uses a qualitative comparative approach to distill different combinations of marketing organizational capabilities and marketing execution capabilities for the presence and absence of a high market, customer, and financial performance. While in the following this study touches upon the critical implication of the findings (concerning individual mechanisms), the primary objective is to further elaborate theory on configurations of marketing organizational capabilities, marketing execution capabilities, and structural contingencies.

The first main findings pertain to the solution set for high market performance and high-profit margin. Past research shows that as market share increases, firms are likely to experience declining marketing costs as a percentage of sales, higher economic rents, shrinking purchase-to-sales ratio, lowering costs, or getting extra concessions form channel members (Demsetz 1973). This experience is because firms use market research extensively, constantly launch new products, and offer high-quality supplementary services, while increasing their advertising activity and strengthening long-lasting relationships with key customers (Slater and Narver 1998). However, these market positions are the supreme area of large corporations as these firms have enough resources to face more significant competitors face-to-face (Hall 1980), market a more extensive product portfolio (Prajogo et al. 2013), and may compete in several markets (simultaneously) while selling in high volumes (Cohn and Lindberg 1974). A fuzzy set analysis shows that for high market performance and high-profit margin, firms require an extensive deployment of marketing organizational capabilities and marketing execution capabilities. These theoretical considerations and findings lead to the following proposition. Proposition 1: The presence of inside-out capabilities, spanning capabilities, advertising capability execution, market capability execution, customer relationship capability execution, and large size leads to high market performance and high-profit margin.

The practical implementation of planned marketing strategy is key to linking marketing efforts with performance (Olson et al. 2005; Vorhies and Morgan 2003). Effective implementation can be tricky, as it requires the coordination of different marketing organizational capabilities and capabilities supporting the implementation/execution of necessary marketing activities (Olson et al. 2005). Also, if planned marketing strategy gets implemented with failure (e.g., ineffectual resource allocation, lack of activities supporting capability implementation or omitting important (or necessary) capabilities from process implementation), firms may not be able to organize their marketing activities properly and will perform poorly (Ruekert and Walker Jr 1987). Woo and Cooper (1981) contend that ineffective low-share businesses follow competitive strategies of effective high-share businesses—with a broad product line, high engagement in R&D activities, insufficient scale of operation to support all activities—that lead to weak market performance and low profitability. The truth table algorithm shows that marketing organizational capabilities or marketing execution capabilities alone is insufficient for low market performance and high-profit margin and that the absence of inside-out capabilities leads to low market performance and high-profit margin. Therefore, this study offers and tests proposition 2. Proposition 2a: The presence of marketing organizational capabilities and the absence of marketing execution capabilities lead to low market performance and low-profit margin. Proposition 2b: The absence of inside-out capabilities leads to low market performance and low-profit margin (no matter how firms are wealthy on other marketing organizational capabilities and activities supporting the execution of marketing capabilities).

Market share building strategies focus on (1) growing market segments, (2) boosting a broad and differentiated product portfolio, (3) encouraging more firms to enter the market, and (4) creating capabilities that may be difficult to imitate by competitors (Barney 1991; Narver and Slater 1990; Nelson and Winter 1982). However, pumping out new product versions and expanding market share may bring short-term gains, but this often comes at the expense of long-term profits (Quelch and Kenny 1994). As the PIMS study shows, there is an optimal market share for firms, and the deviation from this optimal market share may reduce firms' profitability (Bloom and Kotler 1975). The fuzzy set analysis shows that a fast deployment of marketing organizational capabilities and activities supporting the execution of marketing capabilities increases market performance but reduces profitability. This argument supports the following proposition. Proposition 3: The presence of inside-out capabilities, spanning capabilities, outside-in capabilities, advertising capability execution, market research capability execution, customer relationship capability execution, highly competitive market environment, and service orientation leads to high market performance and low-profit margin.

Firms with innovative new products that face a low level of competition can earn high profits, which attracts other firms to enter the market/market segment, and the increased level of competition will ultimately erode high-profit levels (Schumpeter 1950). However, as Schumpeter (1950) contends, firms with younger product portfolios with a relative monopoly market position and stable environments can still be profitable. Notably, the fuzzy set results show that firms with robust new product/service development/introduction capabilities that are strong at advertising and market research, are service oriented, have a large size, and operate in stable environments can attain high profits (despite having low sales volume and market share). This perspective leads to the following proposition. Proposition 4: The presence of spanning capabilities, advertising capability execution, market research capability execution, stable environments, large size, and service orientation leads to low market performance and high-profit margin.

Satisfied customers are more loyal to the firm and stay longer with the company (Fornell 1992), which generates a steady stream of future cash flow (Reichheld and Sasser Jr 1990), which should be reflected in the firm's asset value generated by its custom base (Anderson et al. 1994). A satisfied consumer base requires an efficient human resource management (Day 1994), continuous new product/service development (Day 1994; Javalgi et al. 2006), an efficient market research capability, and strong customer relationship management (Jayachandran et al. 2004). Findings of the fuzzy set analysis show that firms with robust inside-out capabilities, strong market research capability execution, and strong customer relationship capability execution—operating in a highly turbulent environment—can reach high customer performance and high-profit rate. This perspective supports the following proposition. Proposition 5: The presence of inside-out capabilities, market research capability execution, customer relationship capability execution, and dynamic environmental forces leads to high customer performance and high-profit margin.

The implementation of planned marketing strategy is crucial to linking marketing efforts with performance (Olson et al. 2005; Vorhies and Morgan 2003). A firm's implementation capabilities may be "partial" as, e.g., a firm may be well equipped to launch a product with reasonable technical services but not being able to provide other components of the strategy as high-quality customer service or an effective marketing communication (Cespedes and Piercy 1996; Piercy 1998). The truth table algorithm shows that firms that are lacking activities that support marketing execution capabilities are likely to experience low customer performance and low-profit margin. Alternatively, a firm's implementation capabilities can be "latent" as, e.g., a firm might possess the technical and human resources required by marketing strategy implementation, but it may lack the ability to deploy resources and leverage on excellent manufacturing ability (Cespedes and Piercy 1996; Piercy 1998). The fuzzy set analysis shows that in the absence of inside-out capabilities, firms perform poorly compared to leading competitors concerning market performance and profit margin. These findings lead to the following formal propositions. Proposition 6a: The presence of marketing organizational capabilities and the absence of marketing execution capabilities lead to low customer performance and low-profit margin. Proposition 6b: The absence of inside-out capabilities leads to low customer performance and low-profit margin (no matter how wealthy firms are on other marketing organizational capabilities and activities supporting marketing execution capabilities).

Increasing customer satisfaction consumes a firm's resources and may also be subject to diminishing returns (Anderson and Mittal 2000; Rust et al. 1995), as increasing customer satisfaction requires products and services with more and better features, available in a greater variety of combinations, or more attention to customers by employees (Anderson and Mittal 2000; Mittal et al. 2005). Thus, revenue from satisfied and loyal customers comes at a cost, and customers may only become profitable over time (Anderson and Mittal 2000). The truth table algorithm shows that building strong relationships with customers through tailor-made services may have a positive effect on customer satisfaction and customer loyalty, but significantly reduces profitability. This discussion supports proposition 7. Proposition 7: The presence of customer relationship capability execution (in addition to being large/small, competing in highly competitive/stable environments, or possessing a wide range of necessary marketing organizational capabilities and activities supporting marketing execution capabilities) leads to high customer performance and low-profit margin.

Firms competing in markets (or industries) with limited competition and specific location advantage can attain high economic returns despite being weak on customer satisfaction (Anderson et al. 1997). For these firms, satisfied customers provide fewer benefits. This occurs when the firm is competing in a highly concentrated market (or industry), where one firm (or a small set of firms) has a stable location advantage, giving the firm a degree of monopoly power that allows the firm to earn high profits (Anderson et al. 1997). Fuzzy set results show that large firms—operating in stable environments—that develop products based on market research data and trying to maintain a close relationship with customers can attain low customer performance and high-profit margin. This perspective supports proposition 8. Proposition 8: The presence of inside-out capabilities, market research capability execution, customer relationship capability execution, and large size leads to low customer performance and high-profit margin.

6 Managerial Implications

A managerial audience can also benefit from the theory and findings in this study. Assessing whether a firm's marketing activities are in line with its strategy and the firm's internal and external contingencies requires the simultaneous assessment of multiple characteristics of the business, and for every strategy, there exists an ideal set of organizational characteristics that leads to superior business performance (or the absence of superior business performance). The managerial relevance of the study can be summarized as follows.

First, in a market-oriented firm, marketing organizational capabilities and resources should be connected as the firm would be able to attract and retain customers and provide them with a value that potential rivals cannot meet. For achieving this outcome, the firm must design and execute work practices (i.e., business and operating processes) that may go beyond activities traditionally considered as the control domain of the marketing function. For this, marketing must be infused as an input to business processes and should contribute to outputs that create and sustain customer value.

Second, capability implementation should flow from strategy formulation that entails the adjustment of organizational structural frameworks and resource deployment around strategic goals (i.e., the manipulation of systems and structures) to attain a specific performance outcome. However, to accomplish marketing capability implementation, individuals in the firm should believe in marketing and customer imperatives, and management priorities should focus on satisfying key customers.

Third, superior business performance may differ along various strategic goals. A sound strategy that rests on strategic offensive moves eventuates in increasing sales volume and market share. In contrary, a customer-focused firm emphasizes the importance of customer satisfaction and loyalty, stimulating the organization to search for ways to better serve customers, but sacrificing short-term profits for long-term customer satisfaction. Because these two objectives do not converge and can often be inversely related, a firm must take necessary trade-off decisions when assigning resources and implementing capabilities.

7 Limitation and Future Research Avenues

Assessing marketing organization fit with marketing strategy goals requires the identification of different configurations for firms seeking to maximize business performance. Although this study outlines and tests a comprehensive theoretical framework combining marketing organizational capabilities, marketing execution capabilities, and contingencies internal and external to the firm to assess how the co-alignment across these factors affects business performance (and the absence of high business performance), it has some notable weaknesses, too. First, this study takes a cross-sectional approach to analysis. A firm may gain or lose the competencies on which a strategy relies on execution. Thus, implementation capabilities change. Future studies might take a dynamic perspective on how implementation capabilities can support strategy execution. Second, some capabilities might not exist in specific organizational culture, thus, making delivery of, e.g., customer satisfaction impossible. Future research could test how different types of organizational cultures might influence capability implementation. Third, capability implementation might link closely with a specific manager who may possess the abilities and power to pursue strategic goals. Future studies could potentially investigate managers' role in capability implementation.

Appendix A Sample Distribution

Area of operations	Frequency	Percent	Firm size (nr. of employees)	Frequency	Percent	Turnover, million HUF	Frequency	Percent	Profit, million HUF	Frequency	Percent
Agriculture	53	13.1	<20	26	6.4	<150	85	21.0	<5	80	19.8
Construction industry	73	18.0	20–49	145	35.8	151–250	62	15.3	6–15	90	22.2
Transport	15	3.7	50–99	105	25.9	251–500	88	21.7	16–50	121	29.9
Wholesale	21	5.2	100–299	81	20.0	501–1000	59	14.6	51–100	49	12.1
Financial services	8	2.0	300–499	20	4.9	>1001	111	27.4	>101	65	16.0
Mining	8	2.0	500–999	16	4.0						
Processing industry	83	20.5	1000–4999	8	2.0						
Telecommunications	8	2.0	>5000	4	1.0						
Retail	37	9.1									
Other services	41	10.1									
Others	58	14.3									
Total	**405**	**100**		**405**	**100**		**405**	**100**		**405**	**100**

Appendix B Measurement Items and Standardized Loadings

Source	Construct	Item	Stand. loading
Adapted from Day (1994)	Inside-out capabilities[a]	1. Effective financial management 2. Efficient human resources management 3. Manufacturing expertise	0.73 0.78 0.65
Adapted from Day (1994) and Theoharakis et al. (2009)	Spanning capabilities[a]	1. New product/service introduction (i.e., successful new product/service introduction) 2. New product/service development (i.e., effective new product/service development)	0.75 0.82
Adapted from Day (1994) and Theoharakis et al. (2009)	Outside-in capabilities[a]	1. Customer linking (i.e., understanding customer changing needs) 2. Channel bonding (i.e., good interpersonal skills with key customers and customer groups) 3. Relationship keeping (i.e., good relationship maintenance with key customers and customer groups)	0.50 0.93 0.91
New scale	Advertising capability execution[b]	1. When advertising products to consumers, we use the media considerably 2. We are present at our fairs and exhibitions of our products 3. We use the internet considerably when advertising	0.63 0.70 0.80
New scale	Market research capability execution[b]	1. We use marketing research considerably 2. Our marketing research is based on the understanding of consumer needs and wants 3. We invest heavily in the creation of well-known, strong brands	0.81 0.78 0.53
New scale	Customer relationship capability execution[b]	1. We put much emphasis on building long-term relationships with our major clients 2. We regularly monitor and analyze the level of actual consumer satisfaction	0.62 0.80
Hooley et al. (2005)	Market performance (subjective)[c]	1. Sales volume achieved 2. Market share achieved	0.92 0.90
Hooley et al. (2005)	Customer performance (subjective)[c]	1. Level of customer satisfaction 2. Level of customer loyalty	0.78 0.77

(continued)

Source	Construct	Item	Stand. loading
–	Turnover (objective)	How much of your company's turnover was in the last financial year, in million forints?	–
–	Profit before taxes (objective)	How much was your company's approximate pre-tax profit for the last financial year, in million forints?	–
Jaworski and Kohli (1993)	Competitive intensity[a]	1. Competition for sales is intense 2. Competition is lively and constantly changing	0.73 0.95
Jaworski and Kohli (1993)	Technological turbulence[a]	1. New products and services are becoming to the market increasingly rapidly 2. Technological changes are becoming faster	0.85 0.55
–	Large firm	Approximately how many employees does the company have?	–
–	Service orientation[d]	Which market is the company most present in?	–

Note: [a]The response options ranged from 1 = strongly disagree to 5 = strongly agree
[b]The response options ranged from 1 = fully disagree to 5 = fully agree
[c]Compared to our main competitors: 1 = much worse, 5 = much better
[d]1 = consumer durables, 2 = consumer goods (FMCG), 3 = basic materials and components, 4 = industrial production equipment, 5 = industrial services, 6 = consumer goods

Appendix C Set Memberships and Calibration Measures/Measures and Sample Descriptives

		Fuzzy set calibration			Measures descriptives		
Condition	Fuzzy set/measure	Fully in	Crossover	Fully out	Median	Max	Min
Inside-out capabilities	Financial management, human resources management, and manufacturing expertise	12.00	9.98	8.00	10.00	15.00	3.00
Spanning capabilities	New product/service development, new product/service introduction	7.00	5.98	5.00	6.00	10.00	2.00
Outside-in capabilities	Customer linking, channel, and relationship keeping	12.00	11.98	9.00	12.00	15.00	6.00

(continued)

		Fuzzy set calibration			Measures descriptives		
Condition	Fuzzy set/measure	Fully in	Crossover	Fully out	Median	Max	Min
Advertising capability execution	Advertising products to achieve high brand awareness and purchase intention	9.00	5.98	4.00	6.00	18.00	3.00
Market research capability execution	Using market research in understanding consumer needs and wants	10.80	7.98	6.00	8.00	18.00	3.00
Customer relationship capability execution	Establishing long-term relationships with customers and monitoring customer satisfaction	9.00	7.98	6.00	8.00	12.00	2.00
Market performance (subjective)	Sales volume and market share	8.00	5.98	5.00	6.00	10.00	2.00
Customer performance (subjective)	Customer satisfaction and customer loyalty	8.00	6.98	6.00	7.00	10.00	4.00
Turnover (objective)	Company's turnover in the last financial year	1300.00	298.00	150.00	400.00	94,000.00	14.00
Profit before taxes (objective)	Company's approximate pre-tax profit for the last financial year	50.00	25.00	5.00	21.00	18,124.00	1.00
Competitive intensity	The degree of competition, including the aggressiveness and the breath of competitive steps	7.00	8.98	10.00	10.00	9.00	2.00
Technological turbulence	The predictability of the technological changes in a given product market	5.00	6.98	8.00	10.00	7.00	2.00
Large firm	Big versus small firms	0.99	–	0.01	0.00	1.00	0.00
Service orientation	Service vs. production orientation	0.99	–	0.01	0.00	1.00	0.00

References

Anderson, E., & Weitz, B. (1989). Determinants of continuity in conventional industrial channel dyads. *Marketing Science, 8*(4), 310–323.

Anderson, E. W., Fornell, C., & Lehmann, D. R. (1994). Customer satisfaction, market share, and profitability: Findings from Sweden. *The Journal of Marketing, 58*(3), 53–65.

Anderson, E. W., Fornell, C., & Rust, R. T. (1997). Customer satisfaction, productivity, and profitability: Differences between goods and services. *Marketing Science, 16*(2), 129–145.

Anderson, E. W., & Mittal, V. (2000). Strengthening the satisfaction-profit chain. *Journal of Service Research, 3*(2), 107–120.

Appiah-Adu, K., & Singh, S. (1998). Customer orientation and performance: A study of SMEs. *Management Decision, 36*(6), 385–394.

Armstrong, J. S., & Overton, T. S. (1977). Estimating nonresponse bias in mail surveys. *Journal of Marketing Research, 14*(3), 396–402.

Barney, J. B. (1991). Firm resources and sustained competitive advantage. Journal of Management (17), 99–120.

Bateson, J. E. G., & Hoffman, K. D. (1999). *Managing services marketing: Text and reading* (4th ed.). Fort Worth, TX: Dryden Press.

Bloom, P. N., & Kotler, P. (1975). Strategies for high market-share companies. *Harvard Business Review, 53*(6), 63–72.

Buzzell, R. D., Gale, B. T., & Sultan, R. G. M. (1975). Market share—A key to profitability. *Harvard Business Review, 53*(1), 97–106.

Buzzell, R. D., & Wiersema, F. D. (1981). Successful share-building strategies. *Harvard Business Review, 59*(1), 135–144.

Carpenter, G. S., & Nakamoto, K. (1989). Consumer preference formation and pioneering advantage. *Journal of Marketing Research, 26* (August), 285–298.

Cespedes, F. V., & Piercy, N. F. (1996). Implementing marketing strategy. *Journal of Marketing Management, 12*(1–3), 135–160.

Cohn, T., & Lindberg, R. A. (1974). *Survival & growth: Management strategies for the small firm.* New York, NY: Amacom.

Collis, D. J. (1994). Research note: How valuable are organizational capabilities? *Strategic Management Journal, 15*(1), 143–152.

Collis, D. J., & Montgomery, C. A. (1998). Creating corporate advantage. *Harvard Business Review, 76*(3), 70–83.

Conner, K. R., & Prahalad, C. K. (1996). A resource-based theory of the firm: Knowledge versus opportunism. *Organization Science, 7*(5), 477–501.

Day, G. S. (1994). The capabilities of market-driven organizations. *The Journal of Marketing, 58*(4), 37–52.

Day, G. S. (1999). Creating a market-driven organization. *Sloan Management Review, 41*(1), 11–12.

Day, G. S. (2011). Closing the marketing capabilities gap. *The Journal of Marketing, 75*(4), 183–195.

Demsetz, H. (1973). Industry structure, market rivalry, and public policy. *The Journal of Law and Economics, 16*(1), 1–9.

Edvardsson, B., Johnson, M. D., Gustafsson, A., & Strandvik, T. (2000). The effects of satisfaction and loyalty on profits and growth: Products versus services. *Total Quality Management, 11*(7), 917–927.

Fiss, P. C. (2007). A set-theoretic approach to organizational configurations. *Academy of Management Review, 32*(4), 1180–1198.

Fiss, P. C. (2011). Building better causal theories: A fuzzy set approach to typologies in organization research. *Academy of Management Journal, 54*(2), 393–420.

Fornell, C. (1992). A national customer satisfaction barometer: The Swedish experience. *The Journal of Marketing, 56*(1), 6–21.

Fornell, C., & Larcker, D. F. (1981). Evaluating structural equation models with unobservable variables and measurement error. *Journal of Marketing Research, 18*(1), 39–50.

Glazer, R. (1991). Marketing in an information-intensive environment: Strategic implications of knowledge as an asset. *The Journal of Marketing, 55*(4), 1–19.

Gresov, C., & Drazin, R. (1997). Equifinality: Functional equivalence in organization design. *Academy of Management Review, 22*(2), 403–428.

Grönroos, C. (1990). *Service management and marketing: Managing the moments of truth in service competition*. Lexington, MA: Lexington Books.

Hair, J. F., Black, W. C., Babin, B. J., Anderson, R. E., & Tatham, R. L. (2006). *Multivariate data analysis* (6th ed.). Uppersaddle River: Pearson Prentice Hall.

Hall, W. K. (1980). Survival strategies in a hostile environment. *Harvard Business Review, 58*(5), 75–85.

Hooley, G. J., Greenley, G. E., Cadogan, J. W., & Fahy, J. (2005). The performance impact of marketing resources. *Journal of Business Research, 58*, 18–27.

Hubbard, R. (2015). *Corrupt research: The case for reconceptualizing empirical management and social science*. Thousand Oaks, CA: SAGE Publications.

Javalgi, R. (R.) G., Martin, C. L., & Young, R. B. (2006). Marketing research, market orientation, and customer relationship management: A framework and implications for service providers. *Journal of Services Marketing, 20*(1), 12–23.

Jaworski, B. J., & Kohli, A. K. (1993). Market orientation: Antecedents and consequences. *The Journal of Marketing, 57*(3), 53–70.

Jayachandran, S., Hewett, K., & Kaufman, P. (2004). Customer response capability in a sense-and-respond era: The role of customer knowledge process. *Journal of the Academy of Marketing Science, 32*(3), 219–233.

Johnson, M. D., Herrmann, A., & Gustafsson, A. (1999). *Customer satisfaction across industries, countries, and time: A test of Alderson's theory of market behavior*. Working paper, University of Michigan Business School, Ann Arbor.

Keller, K. L. (1993). Conceptualizing, measuring, and managing customer-based brand equity. *The Journal of Marketing, 57*(1), 1–22.

Lee, J., Lee, J., & Feick, L. (2001), "The impact of switching costs on the customer satisfaction-loyalty link: mobile phone service in France", Journal of Services Marketing, 15, 35–48.

Leonard, D., & Rayport, J. F. (1997). Spark innovation through empathic design. *Harvard Business Review, 75*(6), 102–113.

Lindell, M. K., & Whitney, D. J. (2001). Accounting for common method variance in cross-selectional research designs. *Journal of Applied Psychology, 86*(1), 114–121.

Martin, J. A., & Eisenhardt, K. M. (2010). Rewiring: Cross-business-unit collaborations in multi-business organizations. *Academy of Management Journal, 53*(2), 265–301.

McKee, Daryl O., P. Rajan Varadarajan, and William M. Pride (1989). Strategic adaptability and firm performance: A Market-contingent perspective. *Journal of Marketing, 53*, 21–35.

Milgrom, P., Qian, Y., & Roberts, J. (1991). Complementarities, momentum, and the evolution of modern manufacturing. *The American Economic Review, 81*(2), 84–88.

Mittal, V., Anderson, E. W., Sayrak, A., & Tadikamalla, P. (2005). Dual emphasis and the long-term financial impact of customer satisfaction. *Marketing Science, 24*(4), 544–555.

Morgan, N. A., Clark, B. H., & Gooner, R. (2002). Marketing productivity, marketing audits, and systems for marketing performance assessment: Integrating multiple perspectives. *Journal of Business Research, 55*(5), 363–375.

Nagy, G., Megehee, C. M., Woodside, A. G., Laukkanen, T., Hirvonen, S., & Reijonen, H. (2017). Achieving requisite variety in modeling firms' strategy heterogeneities: Explaining paradoxical firm-market performances. *Industrial Marketing Management, 6*, 100–128.

Nagy, G., Megehee, C.M., & Woodside, A.G. (2018). Computing with words. In Woodside, A. (2018) (Ed.), Improving the marriage of modeling and theory for accurate forecasts of outcomes (Advances in Business Marketing and Purchasing, Vol. 25), Emerald Publishing Limited, 155–236.

Narver, J. C., & Slater, S. F. (1990). The effect of a market orientation on business profitability. *The Journal of Marketing, 54*(4), 20–35.
Nelson, R. R., & Winter, S. G. (1982). *An evolutionary theory of economic change*. Cambridge, MA: Harvard University Press.
Noble, C. H., & Mokwa, M. P. (1999). Implementing marketing strategies: Developing and testing a managerial theory. *The Journal of Marketing, 63*(4), 57–73.
Olson, E. M., Slater, S. F., Tomas, G., & Hult, M. (2005). The importance of structure and process to strategy implementation. *Business Horizons, 48*(1), 47–54.
Pelham, A. M., & Wilson, D. T. (1996). A longitudinal study of the impact of market structure, firm structure, strategy, and market orientation culture on dimensions of small-firm performance. *Journal of the Academy of Marketing Science, 24*(1), 27–43.
Phillips, L. W., Chang, D. R., & Buzzell, R. D. (1983). Product quality, cost position and business performance: A test of some key hypotheses. *The Journal of Marketing, 47*(2), 26–43.
Piercy, N. F. (1998). Marketing implementation: The implications of marketing paradigm weakness for the strategy execution process. *Journal of the Academy of Marketing Science, 26*(3), 222–236.
Podsakoff, P. M., MacKenzie, S. B., Lee, J. Y., & Podsakoff, N. P. (2003). Common method biases in behavioral research: A critical review of the literature and recommended remedies. *Journal of Applied Psychology, 88*(5), 879–903.
Powell, T. C., Lovallo, D., & Fox, C. R. (2011). Behavioral strategy. *Strategic Management Journal, 32*(13), 1369–1386.
Prajogo, D. I., McDermott, C. M., & McDermott, M. A. (2013). Innovation orientations and their effects on business performance: Contrasting small- and medium-sized service firms. *R&D Management, 43*(5), 486–500.
Quelch, J. A., & Kenny, D. (1994). Extend profits, not product lines. In *Make sure all your products are profitable* (2nd ed.), Harvard business review on point collection (pp. 14–24).
Ragin, C. C. (2008). *Redesigning social inquiry: Fuzzy sets and beyond*. Chicago, IL: University of Chicago Press.
Ratnatunga, J., & Ewing, M. T. (2005). The brand capability value of integrated marketing communication (IMC). *Journal of Advertising, 34*(4), 25–40.
Reichheld, F. F. (1993). Loyalty-based management. *Harvard Business Review, 71*(2), 64–73.
Reichheld, F. F., & Sasser, W. E., Jr. (1990). Zero defections: Quality comes to services. *Harvard Business Review, 68*(5), 105–111.
Reinartz, W., & Kumar, V. (2002). The mismanagement of customer loyalty. *Harvard Business Review, 80*(7), 86–95.
Roberts, P. W. (1999). Product innovation, product-market competition, and persistent profitability in the U.S. pharmaceutical industry. *Strategic Management Journal, 20*(7), 655–670.
Robertson, T. S., & Gatignon, H. (1986). Competitive effects on technology diffusion. *The Journal of Marketing, 50*(3), 1–12.
Ruekert, R. W., & Walker, O. C., Jr. (1987). Marketing's interaction with other functional units: A conceptual framework and empirical evidence. *The Journal of Marketing, 51*(1), 1–19.
Ruekert, R. W., Walker, O. C., Jr., & Roering, K. J. (1985). The organization of marketing activities: A contingency theory of structure and performance. *The Journal of Marketing, 49*(1), 13–25.
Rust, R. T., Zahorik, A. J., & Keiningham, T. L. (1995). Return on quality (ROQ): Making service quality financially accountable. *The Journal of Marketing, 59*(2), 58–70.
Sashittal, H. C., & Jassawalla, A. R. (2001). Marketing implementation in smaller organizations: Definition, framework, and propositional inventory. *Journal of the Academy of Marketing Science, 29*(1), 45–69.
Schmalensee, R. (1978). A model of advertising and product quality. *Journal of Political Economy, 86*(3), 485–503.
Schumpeter, J. A. (1950). *Capitalism, socialism, and democracy* (3rd ed.). New York, NY: Harper and Row.
Simon, H. A. (1990). Invariants of human behavior. *Annual Review of Psychology, 41*(1), 1–20.

Slater, S. F., & Narver, J. C. (1998). Customer-led and market-oriented: Let's not confuse the two. *Strategic Management Journal, 19*(10), 1001–1006.
Smallwood, D. E., & Conlisk, J. (1979). Product quality in markets where consumers are imperfectly informed. *Quarterly Journal of Economics, 93*(1), 1–23.
Srivastava, R. K., Shervani, T. A., & Fahey, L. (1998). "Market-Based Assets and Shareholder Value: A Framework for Analysis," Journal of Mar keting (62), 2–18.
Szymanski, D. S., Bharadwaj, S. G., & Rajan Varadarajan, P. (1993). An analysis of the market share-profitability relationship. *The Journal of Marketing, 57*(3), 1–18.
Theoharakis, V., Sajtos, L., & Hooley, G. (2009). The strategic role of relational capabilities in the business-to-business service profit chain. *Industrial Marketing Management, 38*(8), 914–924.
Thorpe, E. R., & Morgan, R. E. (2007). In pursuit of the 'ideal approach' to successful marketing strategy implementation. *European Journal of Marketing, 41*(5/6), 659–677.
Urry, J. (2005). The complexity turn. *Theory, Culture & Society, 22*(5), 1–14.
Vorhies, D. W., Harker, M., & Rao, C. P. (1999). The capabilities and performance advantages of market-driven firms. *European Journal of Marketing, 33*(11/12), 1171–1202.
Vorhies, D. W., & Morgan, N. A. (2003). A configuration theory assessment of marketing organization fit with business strategy and its relationship with marketing performance. *The Journal of Marketing, 67*(1), 100–115.
Walker, O., & Ruekert R. (1987). Marketing's role in the implementation of business strategies: a critical review and conceptual framework. *Journal of Mar-keting, 51*(3):15–33.
Woo, C. Y. Y., & Cooper, A. C. (1981). Strategies of effective low share businesses. *Strategic Management Journal, 2*(3), 301–318.
Woodside, A. G. (2013). Moving beyond multiple regression analysis to algorithms: Calling for adoption of a paradigm shift from symmetric to asymmetric thinking in data analysis and crafting theory. *Journal of Business Research, 66*(4), 463–472.
Workman, J. P., Jr., Homburg, C., & Gruner, K. (1998). Marketing organization: An integrative framework of dimensions and determinants. *The Journal of Marketing, 62*(3), 21–41.

Chapter 4
Modeling Human Resource Outcomes

April J. Spivack and Arch G. Woodside

Abstract Who are the knowledge workers perceiving high versus low location autonomy? Do these workers consistently select work environments to enhance their well-being or to enhance their productivity? The study here frames the causal conditions for answering these research questions for case outcomes in response to calls (Misangyi et al. 2017; Woodside 2014) to embrace complexity theory in management research by constructing and testing asymmetric case-based models of decisions and outcomes. Complexity theory includes the tenet that both negative (low) and positive (high) scores for the same antecedent condition may be present in different configurations indicating the same outcome condition (i.e., the equifinality tenet). A second complexity tenet is the causal conditions (i.e., configurations) indicating cases having high outcome scores (e.g., outcome of a "go" decision) may have a few of the same as well as different ingredients than the causal conditions indicating cases having low outcome scores (e.g., the outcome of a "no-go" decision). The present study examines eight propositions relating to knowledge workers' choices of work environments including the following statements. P1: Knowledge workers high in intrinsic work motivation consistently select work environment choices to enhance productivity. P2: Knowledge workers with high scores in perceived location autonomy (PLA) are workers who consistently select work environment choices to enhance well-being and/or work productivity. The study includes examining these two and six additional propositions empirically using a sample of full-time professional knowledge workers. The findings deepen and expand on prior symmetric-based theory and analysis.

Keywords Complexity · Motivation · Perceived location autonomy · Work environment choice

A. J. Spivack (✉)
Coastal Carolina University, Wall College of Business Administration, Conway, SC, USA
e-mail: aspivack@coastal.edu

A. G. Woodside
Yonsei University, Yonsei Frontier Lab, Seoul, Republic of Korea
e-mail: arch.woodside@bc.edu

A. G. Woodside (ed.), *Accurate Case Outcome Modeling*,
https://doi.org/10.1007/978-3-030-26818-3_4

1 Introduction

With the rise of information and communication technologies (ICTs) and their permitting diverse work environments, knowledge workers (KWs) are increasingly experiencing higher levels of perceived location autonomy (PLA) as a result of job re-design to incorporate the use of ICTs in flexible work arrangements. PLA is defined as the autonomy to generate, evaluate, and choose where to perform one's work tasks (Spivack and Milosevic 2018). Hence, research on the antecedents and outcomes of PLA is relevant in human resource management as knowledge is currently lacking on the causal conditions indicating cases (i.e., workers) high versus low in PLA as well as PLA's link to outcomes of work environment choices. The human resource management (HRM) implications of PLA research include knowledge of how workers high versus low in PLA make work environment choices to enhance personal well-being versus enhance work productivity—and how HRM can identify workers seeking to accomplish one or both outcomes as well as the workers seeking neither outcomes (Guest 2017; Zeijen et al. 2018).

The business press literature in the late teen years of the twenty-first century identifies technology-induced dynamics of work environmental contexts as a topic that needs addressing. Haaff (2017, p. 1) proclaims: "Remote workers are outperforming office workers--here's why: research shows that office workers cannot concentrate at their desks." James (2018, p. 1) concludes: "Case closed: work-from-home is the world's smartest management strategy. Research from Stanford and other sources reveals that working from home vastly increases productivity." Fell (2018, p. 1) recommends: "If your company hasn't yet embraced remote work, it's time to pay attention." The study here focuses not on KW productivity increases but on precursor issues to work productivity: building and testing case-based models of identifying KWs high versus low in PLA and how these two conditions influence their choices of work environments that increase their well-being and/or work productivity.

In addition to PLA, intrinsic work motivation is likely to influence work environment choices. As prior research suggests, KWs high in intrinsic work motivation may inordinately seek to support their organizations' productivity objectives (Deci and Ryan 1985); such support is likely to prompt a preference for work environments that enhance work productivity (Spivack and Milosevic 2018). The present study explores possible impacts of KW's with high intrinsic work motivation on their work environment preferences.

This study examines the causal conditions linking PLA, intrinsic work motivation, and other knowledge worker characteristics to work environment choices. Rather than applying the conventional and pervasive variable-directional relationship approach to modeling and testing via null significance hypothesis tests (NHSTs), the study here applies a case-based approach to answer four research questions and examine these models via "somewhat precise outcome tests" (SPOTs) (Hsiao et al. 2015). The four research questions include what are the causal conditions indicating cases high versus low in PLA? Are workers high in PLA consistently high in intrinsic work motivation? What are the causal conditions indicating cases (i.e., workers) high versus low in work environment choices to enhance personal well-being and/or work productivity? Do some workers high in PLA seek both work environment outcomes, or do they perceive the selection to be an either/or option? The present study provides useful answers for understanding the causal antecedent conditions and outcomes of KWs having high versus low preferences for work environments to enhance their well-being and/or work productivity. A possibly unsurprising finding is few KWs high or low in PLA seek the simple outcome condition of a work environment focusing only on increasing work productivity (see Table 4.1).

Following this introduction, Sect. 2 presents a set of propositions of complex antecedent conditions indicating KWs high versus low in PLA as well as how these two outcome conditions influence KW preferences for work environmental contexts. Section 2 includes tenets of complexity theory because the propositions build from a foundation of these tenets. Section 3 describes the method in the study to test the propositions empirically. Section 4 presents the findings. The discussion section in Sect. 5 includes the study's contribution to theory and human resources' practice; limitations and insights for future research in KW's preferences and choices of work environmental contexts conclude.

2 Case-Based Outcome Modeling of Knowledge Workers' Environment Choices

Figure 4.1 is a visual summary of a case-based model of KWs' environment choices. This model builds from a foundation of four complexity theory tenets (Urry 2005; Woodside 2014). The following statements represent these tenets. Tenet 1: A single condition is insufficient consistently for indicating cases having high scores for an outcome condition. Thus, learning that KWs have high PLAs is insufficient for identifying these workers as highly preferring work environments that increase personal well-being. Tenet 2: A few combinations of conditions—complex antecedent conditions consisting of two or more single conditions—are sufficient for an outcome condition. Thus, identifying KWs with high PLA as well as having high intrinsic work motivation prefer work environments that increase work productivity. Tenet 3: The equifinality condition—a few, not just one, complex antecedent condition points to the same outcome. Thus, a few sets of KWs having different identities (i.e., different combinations of attributes in the Venn diagram on the left side in Fig. 4.1) will be likely to have the same work environment preferences. Tenet 4: In different complex antecedent conditions, the same single condition can either support or be a negative influence on the same outcome condition. Thus, different KWs high or low in PLA may prefer work environments that enhance personal well-being—high or low PLA as an ingredient for this outcome depends upon the

Table 4.1 Propositions and summary of the findings

Proposition	Statement supported?	Consistency coverage		
P1a:	Knowledge workers high in intrinsic motivation are high in choices of environment to enhance work productivity	0.61	0.67	No
P1b:	Knowledge workers with low intrinsic motivation are low in choices of environment to enhance work productivity	0.69	0.62	No
P2a:	Knowledge workers with high scores in perceived location autonomy indicates workers high in choices of environment to enhance Well-being AND work productivity	0.51	0.74	No
P2b:	Knowledge workers with high scores in perceived location autonomy AND high scores in intrinsic motivation indicate workers high in choices of work environment to enhance work productivity	0.66	0.50	No
P3a:	Knowledge workers' identity configurations indicate choices of environment to enhance Well-being	0.81	0.21	Yes
P3b:	Knowledge workers' identity configurations indicate choices of environment to enhance work productivity	0.80	0.11	Yes
P3c:	Knowledge workers' identity configurations indicate choices of environment to enhance BOTH Well-being and work productivity	0.83	0.07	Yes
P3d:	Knowledge workers' identity configurations indicate choices of environment to enhance Well-being AND not work productivity	0.78	0.06	Yes
P3e:	Knowledge workers' identity configurations indicate choices of environment to enhance work productivity AND not Well-being	0.81	0.12	Yes
P3f:	Knowledge workers' identity configurations indicate choices of environment neither to enhance Well-being nor to enhance work productivity	0.84	0.07	Yes
P4:	Knowledge workers high in perceived location autonomy indicates workers high in intrinsic work motivation	0.70	0.67	No
P5a:	Unique knowledge workers' identity configurations indicate high perceived location autonomy	0.86	0.06	Yes
P5b:	Unique knowledge workers' identity configurations indicate low perceived location autonomy	0.79	0.28	Yes
P6a:	Unique knowledge workers' identity configurations indicate high intrinsic motivation	0.81	0.15	Yes
P6b:	Unique knowledge workers' identity configurations indicate low intrinsic motivation	0.82	0.18	Yes
P7a:	Unique knowledge workers' configurations including perceived location autonomy indicate high intrinsic motivation	0.83	0.34	Yes
P7b:	Unique knowledge workers' configurations including perceived location autonomy indicate low intrinsic motivation	0.79	0.28	Yes
P8a:	Unique knowledge workers' configurations, including perceived location autonomy and intrinsic motivation, indicate choices of environment to enhance BOTH Well-being and work productivity	0.83	0.20	Yes
P8b:	Unique knowledge workers' configurations, including perceived location autonomy and intrinsic motivation, indicate choices of environment to enhance Well-being AND not work productivity	0.88	0.11	Yes
P8c:	Unique knowledge workers' configurations, including perceived location autonomy and intrinsic motivation, indicate choices of environment to enhance NEITHER Well-being NOR work productivity	0.85	0.23	Yes

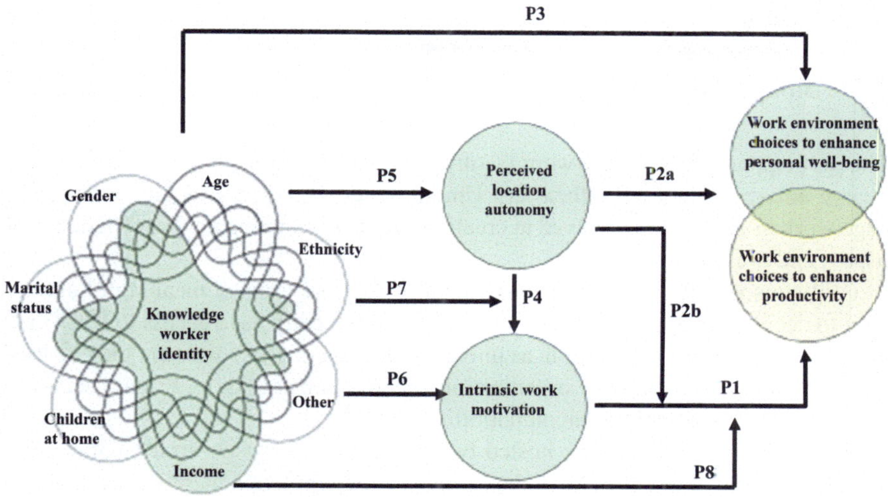

Fig. 4.1 Case-based outcome models of work environment choices

additional conditions among the ingredients in alternative complex antecedent conditions leading to the same outcome. Reflecting these tenets, eight sets of propositions diagrammatically appear in Fig. 4.1. For testing the third proposition (P3) appearing in Fig. 1, specific model propositions A and B were set ex ante to study's empirical analyses. Model A: high-income, unmarried, no children, young, male → choice of CEW (CEW = choice of work environment to enhance personal well-being). Rationale: Possibly model A describes care-free individuals having few responsibilities seeking to enjoy life's human pleasures—the findings do not support model A possibly due to the fact that very few cases fit this model's description. Model B: high-income, married, children, young, female → CEW—the findings do not support B, again likely due in part that few cases fit model B's description. Unsurprisingly, given the absence of prior configurational modeling on workers' work environment choices and the tenets of complexity theory, specifying ex ante as to which specific combinations of antecedent conditions indicate CEW accurately is a difficult task.

Among identity characteristics, the study here includes gender as prior research demonstrates that there are gender differences in perceptions of autonomy in knowledge-intensive firms (Truss et al. 2012). Similarly, the study includes age in response to prior research that indicates age can influence the pattern of engagement with flexible work options (Loretto and Vickerstaff 2015). Furthermore, this study includes other identity characteristics that could reflect resources available to workers, such as income and ethnicity, as well as potential sources of non-work demands, such as marital status and having children under age 18 living at home (Grzywacz and Marks 2000; Tichenor 1999).

3 Method

3.1 Sample and Procedures

The participants of this study were faculty members of the academic community at a large university in the Southeastern United States. University faculty are knowledge workers as they are involved in creating and disseminating new knowledge, via papers and presentations.

(Williamson and Cable 2003). Additionally, they are granted meaningful autonomy to decide where their work is completed and typically have access to ICTs that afford work mobility. In addition, as university members, these workers may be less subject to a variety of other organization-specific constraints (e.g., data security risks, organizational culture incompatibility).

A link to the online survey, hosted by QuestionPro, was sent to a random subsample of faculty ($n = 540$), out of the campus-wide population of 980 faculty. Upon clicking the link to access the survey, respondents had the opportunity to agree to participate after reading an informed consent statement. The measures used in this study were from published scales. From the 540 faculty solicited through an email invitation to participate in the study, there were 184 usable responses, yielding a 34% response rate, which is much higher than the 10% response rate typical of web-based surveys today (Mol 2017).

The sample was composed of 46% male and 54% female respondents. Age of the participants was split across categories: 1% 18–24 years old, 6% 25–34 years old, 29% 35–44 years old, 27% 45–54 years old, and 35% aged 55+. Ethnicity was predominantly white/Caucasian (88%), with 2% being Asian, 5% black or African American, 2% Hispanic, or Latino, and 3% others. Marital status was 17% single, 69% married, 8% divorced, 1% widowed, and 5% cohabitating. Dependent children lived with 42% of the sample. Yearly household income was largely above $90,000 with 54% of the sample in that group, while 1% earned less than $18,000, 5% earned between $25,000 and $45,000, 12% earned between $45,000 and $60,000, and 29% earned between $60,000 and $90,000.

3.2 Measures

Perceived Location Autonomy PLA was measured using seven items with a 5-point Likert-type scale from Spivack and Milosevic (2018) to reflect feeling free to decide where to work. The Cronbach alpha of the scale was 0.85. Sample items included the following: "I have the freedom to decide where to complete my work," "It is basically my own responsibility to find or create an environment that allows me to get my work done," and "I feel free to work off-site."

Intrinsic Work Motivation Intrinsic work motivation was measured using the Work Extrinsic and Intrinsic Motivation Scale (Tremblay et al. 2009), including three items for each of six subscales on a 7-point scale: intrinsic motivation, integrated regulation, identified regulation, introjected regulation, external regulation, and motivation. The Cronbach alpha of each subscale was 0.78, 0.79, 0.67, 0.78, 0.76, and 0.80, respectively. Sample items included "Because I chose this type of work to attain my career goals" and "Because it allows me to earn money." Each subscale score was used to create an overall index of intrinsic motivation, called the "work self-determination index," following the procedure used by Tremblay et al. (2009); the subscales were multiplied by weight factors. The controlled subscales, referring to external motivation, are weighted negatively, and the intrinsic subscales, referring to internal motivation, are weighted positively. The more controlled the regulatory style represented by a subscale, the larger is its negative weight; and the more intrinsic the regulatory style represented by a subscale, the larger is its positive weight. The overall intrinsic motivation score was computed using the following formula: ((3 × intrinsic motivation) + (2 × integrated motivation) + identified regulation) – ((introjected regulation) + (2 × external regulation) + (3 × amotivation)).

Environment Choices Choice of environments to enhance worker well-being was measured using the scale developed by Spivack and Milosevic (2018). Sample items reflecting choice of environments enhancing worker well-being included the following: "The environments I work in contribute positively to my work life satisfaction," "The environments I work in contribute positively to my work–life balance," and "The environments I work in make me feel mentally healthy." Similarly, choice of environments to enhance worker productivity was measured by the scale developed by Spivack and Milosevic (2018). Sample items reflecting choice of environments enhancing productivity included the following: "This environment is the optimal environment for these tasks," "This environment puts me in the right frame of mind to work on these tasks," and "After working in this environment, I usually feel that I have been productive." The responses for each scale were averaged; Cronbach's alpha was 0.93 and 0.91, respectively, for the two scales.

3.3 Analysis

Data analyses include four-corner modeling and asymmetric algorithm construction and testing using the software program fsQCA.com (fsQCA "fuzzy-set qualitative comparative analysis"). Four-corner modeling includes constructing causal configurations of complex antecedent statements for cases in each set of cases in the four corners of cross-tabs of quintiles or deciles of cases for two or more variables (Feurer et al. 2016; Nagy et al. 2017; Woodside 2014). With sample sizes reasonably large (e.g., $n > 50$), discretizing (particularly useful to do by quintiles) two continuous variables and cross-tabulating the two sets of cases result in a few-to-many cases appearing in all 25 cells—if using quintiles—even when the main effect size between the two variables is large ($r2 \geq 0.25$). As McClelland (1998) indicated, converting continuous variables into quintiles is more than just a data manipulation exercise; such conversions represent a shift from variable-focused to case-focused research. A researcher can expect to find a combined 10–20% of the cases that display contrarian associations to variable-directional hypotheses and confirmed directional associations, while the vast majority of studies using the current symmetric-testing dominant logic ignore the occurrences of these cases.

Testing by fsQCA requires calibration of all variable scales into calibrated scales, with scores ranging from end points of 0.00 for full non-membership to 1.00 for full membership. The fsQCA program provides calibrated scores. Fuzzy-set calibration makes use of external information on the degree to which cases satisfy membership criteria and not inductively derived determination criteria (e.g., not using sample means). To perform fuzzy-set calibration, criteria are necessary for three breakpoints—0.05 for threshold for full non-membership, 0.50 for the crossover point of maximum membership ambiguity, and 0.95 for the threshold of full membership. Specifying the original values for these three breakpoints permits the software to calibrate all remaining scores. Calibrated scores are membership scores and not probabilities. For the present report, the following simple procedure was used to calibrate the original multiple-value scales. Scores for cases in the highest quintile equal 0.95, calibrated scores for cases in the middle quintile were set at 0.50, and calibrated scores for cases in the lowest quintile were set at 0.05.

In fsQCA, the consistency and coverage indexes are the metrics indicating the usefulness of a given model of a simple antecedent condition or a set of complex antecedent conditions for predicting scores in an outcome condition. The consistency index gauges the degree to which the cases share the antecedent condition (simple or complex) in displaying the outcome in question—consistency is analogous to correlation or "*r*" in statistical analysis. From Ragin (2008, p. 52), fuzzy set-theoretic consistency can be calculated by "Consistency ($X_i \leq Y_i$) = Σ [min (X_i, Y_i)]/Σ (X_i), where min indicates the selection of the lower of the two values. When the X_i values are all less than or equal to their corresponding Y_i values, the consistency score is 1.0; when there are only a few near misses, the score is slightly less than 1.0; when there are many inconsistent scores, with some X_i values greatly exceeding their corresponding Y_i values, consistency may drop below 0.5."

The coverage index in fsQCA assesses the degree to which a simple or complex antecedent condition (recipe) "accounts for" instances of an outcome condition—coverage is analogous to "R^2" in statistical analysis. The score for a complex antecedent condition is equal to the lowest score for the simple antecedent conditions within the complex statement—this lowest score indicates the shared membership the simple conditions have within the configural statement. As per Ragin (2008, p. 57), "the measure of fuzzy-set coverage is simply the overlap expressed as a proportion of the sum of the membership scores in the outcome": Coverage ($X_i \leq Y_i$) = Σ [min (X_i, Y_i)]/Σ (Y_i). In short, the formula for coverage of Y by X substitutes Σ (Y_i) for Σ (X_i) in the denominator of the formula for consistency.

Ex ante, the study's objective was set such that an fsQCA model is useful when its consistency is equal to or above 0.82 and its coverage is above 0.00. In other words, a consistency equal to or above 0.82 had to be achieved for concluding that a proposition was fully supported by the findings. Coverage indicates the share of cases that are high in the outcome condition that the proposition represents empirically. Coverage can be very low, but a model is still useful if the consistency is high. A high consistency score (e.g., consistency = 0.82) indicates high membership scores in the outcome condition for nearly all cases with high scores in the simple or complex antecedent condition, with most other cases fitting an asymmetric sufficiency distribution (i.e., Panel C in Fig. 4.2). See Ragin (2008) for detailed training on using fsQCA.

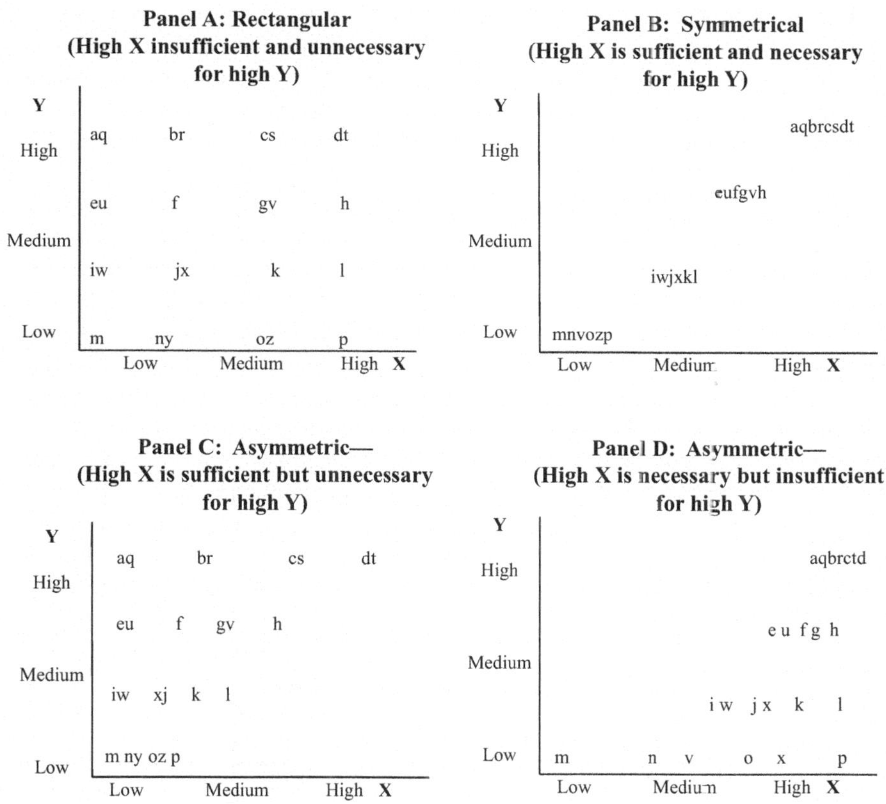

Fig. 4.2 Hypothetical relationships where X is a complex configural condition (e.g., workers high in intrinsic work motivation as X indicates workers high in work environment choice to increase productivity as Y). *Notes*: Panel A: Cases high in X are both high and low in Y. Findings are not symmetric nor asymmetric. Panel B illustrates asymmetric theory and findings. Regression analysis and structural equation modeling (SEM) illustrate symmetric model construction and testing. Panel C: Rather than the pervasive use of symmetric theory and analysis (i.e., constructing multiple regression models and the use of ANOVA tests of null hypothesis, the foundation for the study here is asymmetric theory and analysis (i.e., model construction to indicate cases having high scores in an outcome condition). The present study specifies and tests for models that illustrate the pattern of cases appearing in panel C. Asymmetric modeling are not attempts to predict both low and high scores in an outcome condition (Y) but only predict high scores in the outcome condition (Y). Panel D: Some antecedent conditions are necessary but insufficient for indicating cases with high scores in Y (e.g., being "male" is a necessary but insufficient condition to indicate players (high Y) playing football in the US National Football League

4 Findings

Table 4.1 includes statements of the study's propositions as well as "consistency" and "coverage" indices and whether or not the proposition was supported (i.e., had fsQCA models with consistency above 0.82 and a coverage over 0.00). Interestingly, the findings do not support propositions 1, 2, and 4 because of low consistency indexes. The XY plots of the relationships between intrinsic work motivation, PLA, and work environment choices represented in these propositions confirmed that contrarian cases occur frequently enough to cause lower consistency scores and also signal that these qualities of cases are insufficient for indicating cases reflecting the proposed outcomes. When knowledge workers' identity configurations are included in propositions, with the indicator variables of PLA, intrinsic work motivation, and choices of work environment, the propositions are supported (see propositions 3, 5, 6, 7, and 8).

Four-corner modeling reveals additional insight into the distribution of cases in cross-tabs by quintile for scores on outcome variables of choices of work environment to enhance productivity by choices of work environment to enhance personal well-being (Table 4.2). Only 20% of cases made choices of work environments to enhance both productivity and well-being, and almost one third of the cases made choices of work environments that did not enhance personal well-being nor productivity, while 10% chose work environments to enhance well-being but not productivity, and only 7% chose work environments to enhance productivity and not personal well-being. While most cases lie in the diagonal conditions as expected, it is interesting that the majority of cases are in the quadrant representing choices of work environments to neither enhance work productivity nor personal well-being.

Table 4.2 Study participants by choice to enhance well-being (CEW) and/or to enhance work productivity. Quintiles: 1 = very low, 5 = very high

~CEW•~CEP (23%) ~CEW•CEP (2%)

Choice to enhance work productivity

		1.00	2.00	3.00	4.00	5.00	
	1.00	8	8	2	1	2	21
Choice to	2.00	10	15	6	1	0	32
enhance	3.00	1	0	2	1	0	4
well-being	4.00	6	11	18	4	4	43
	5.00	11	11	21	21	20	84
	Total	36	45	49	28	26	184

CEW•~CEP (21%) CEW•CEP (27%)

Note. Example, the largest (27%) share of respondents selected both choices to enhance well-being and to enhance work productivity. The smallest share (2%) selected only choices to enhance work productivity

Four-corner modeling also highlighted the distribution of cases in cross-tabs by quintile for scores on PLA by scores on both outcome variables of choices of work environment to enhance productivity and choices of work environment to enhance personal well-being (Tables 4.3a and 4.3b). Again, the highest number of cases were expected on the diagonal of high PLA and high CEW (choice of work environment to enhance personal well-being) and high PLA and high CEP (choice of work environment to enhance work productivity). There are unexpected cases in the other quadrants, however.

Together, the four-corner modeling results indicate that the recipes to achieve the desired outcome scores are not a simple linear or symmetric relationship and there is reason to pursue fsQCA. The additional insight from fsQCA will help to identify what conditions indicate both the expected and these unusual outcome patterns among cases in the data. A deeper dive to analysis by algorithms considers causal recipes that do provide high consistency of cases in all four corners of the tables.

Table 4.3a Study participants by perceived location autonomy (PLA) and/or choice of work environment to enhance well-being (CEW). Quintiles: 1 = very low, 5 = very high

a

~PLA•~CEW ~PLA•CEW

PLA \ CEW	1.00	2.00	3.00	4.00	5.00	Total
1.00	16	6	0	2	6	30
	29%			10%		
2.00	12	17	1	6	4	40
3.00	6	20	5	11	11	53
4.00	5	8	2	8	9	32
	13%			19%		
5.00	4	6	1	10	8	29
Total	43	57	9	37	38	184

PLA•~CEW PLA•CEW

Phi = 0.45, p < 0.002

Table 4.3b Study participants by perceived location autonomy (PLA) and/or choice of work environment to enhance work productivity (CEP). Quintiles: 1 = very low, 5 = very high

b

~PLA•~CEP ~PLA•CEP

PLA \ CEP	1.00	2.00	3.00	4.00	5.00	Total
1.00	11	9	4	3	3	30
	23%			7%		
2.00	11	11	11	4	3	40
3.00	8	13	17	10	5	53
4.00	3	8	10	6	5	32
	10%			14%		
5.00	3	4	7	5	10	29
Total	36	45	49	28	26	184

PLA•~CEP PLA•CEP

Phi = 0.384, p < 04

Table 4.4 Testing P3a: Case-based models of respondents with high membership scores in choice to enhance well-being. Key: CEW = choice of environment to enhance well-being

P3: Model: cew = f(inc, ethnicity, marital, child_home, age, gender)		
Possible theoretical configurations: 3 income levels X 2 ethnic groups X 2 marital status X 2 child status X 3 age X 2 gender = 144 combinations.		
Specific model propositions based on conjecture:		
A: high-income, non-white, unmarried, no children, young, male → choice of CEW (findings do not support A).		
B. High-income, white, married, children, young, female → CEW (findings do not support B).		
Frequency cutoff: 2.00; consistency cutoff: 0.81		
Models	**Coverage**	**Consistency**
1 ~inc•white•marital•~age•gender	0.05	0.79
2 ~inc•white•marital•child_home•gender	0.06	0.81
3 inc•~white•marital•~child_home•~age•~gender	0.03	0.93
4 inc•white•marital•~child_home•age•~gender	0.11	0.84
5 inc•white•~marital•~child_home•age•gender	0.05	0.81
Solution coverage: 0.21; solution consistency: 0.81.		
Examples of how to read models in these findings: Model 1 includes young, married, white, males with low income both with and without children at home Model 2 includes both young/old, married, white males with low income and children at home Model 3 includes young, married, nonwhite, females with high income and no children at home Model 4 includes older, married, white, females with high income and no children at home Model 5 includes older, unmarried, white, males with high income, and no children at home		

An example of the insights gained through fsQCA is shown in Table 4.4 as well as how to interpret the models generated. Specifically, a mix of configurations of 6 conditions yield 5 case-based models out of 144 possible combinations represented in the high CEW quadrant shown. To understand how to interpret the models, please see Table 4.4.

The conclusion of the four-corner modeling results is that the diagonals receive insufficient support—high CEW does not alone predict high CEP, nor does high PLA alone predict high CEW or CEP. Instead, high PLA may be a contributing factor to high CEW and/or CEP in some configurations with additional antecedents—the findings below do support this perspective.

4.1 Findings for Models for High CEW and CEP

Table 4.5 presents findings for propositions 8a, b, and c of the study that appear in Table 4.1. Table 4.5a presents three complex antecedent models that include knowledge worker identity factors, PLA, and intrinsic work motivation that are associated

Table 4.5a Testing P8a: knowledge worker identity, perceived location autonomy, and intrinsic work motivation indicate workers high in both CEW AND CEP

	Coverage	Consistency
1. Locauto•motv•~inc•white•~child_home•age•~gender	0.10	0.88
2. Locauto•motv•~inc•white•marital•~child_home•age	0.08	0.85
3. ~locauto•motv•inc•white•marital•child_home•age•gender	0.10	0.85

P8a: model—cew_cep = f(locauto, motv, inc, ethnicity, marital, child_home, age, gender); frequency cutoff, 2.00; consistency cutoff, 0.83

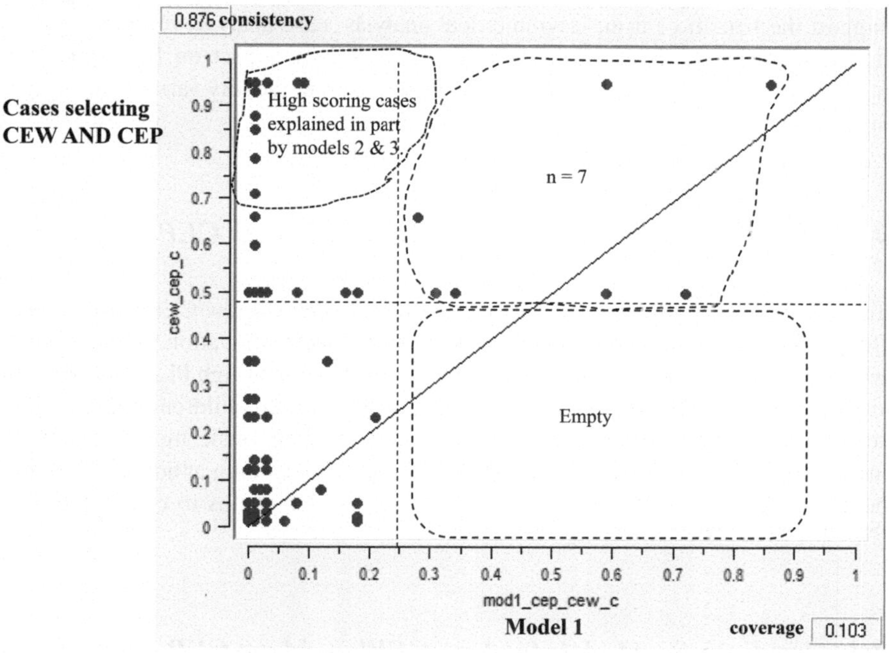

Fig. 4.3 XY plot for model 1 indicating cases with high scores for both CEW AND CEP. Notes: model 1— older, white, females with low income and no children at home, high in perceived location autonomy, and high in intrinsic work motivation

with outcomes high in choices of work environment to enhance both personal well-being and work productivity. Figure 4.3 presents an XY plot of model 1 for the high CEW and high CEP quadrant. Model 1 has high consistency of 0.88 and represents seven individuals in the data that match the specific configuration of older, white females with low income and no children at home, being high in perceived location autonomy and high in intrinsic work motivation.

Model 2 includes high PLA, high intrinsic motivation, low income, older, and white married males and females with no children at home. Model 3 includes low PLA, high intrinsic motivation, high income, and older white males that are married with children at home. All three models include knowledge workers high in intrinsic work motivation and possessing identity characteristics of older and white, but they vary in the inclusion of PLA, marital status, gender, and whether children are at home. As a result, the three models indicate three different "complex causal conditions" (Ragin 1999) associating with high CEW and CEP.

These findings complement and expand on the findings in the literature focusing on gender and perceptions of autonomy among knowledge workers (Truss et al. 2012) and gender, age, and the use of flexible work arrangements (Loretto and Vickerstaff 2015). This body of work expands and deepens the field via configurational modeling knowledge worker identities, PLA, intrinsic work motivation, and outcomes of CEW and CEP. The configurational findings in the present study also suggest the benefit of using asymmetrical analysis; case analysis revealed hidden linkages in Spivack and Milosevic's (2018) work—an impact on PLA, intrinsic motivation, CEW, and CEP related to knowledge worker identity variables (i.e., age and gender).

4.2 *Findings for Models for High CEW and Low CEP*

Table 4.5b presents two models that predict high CEW and low CEP consistently. Both models in Table 4.5b include high intrinsic work motivation, younger, and white knowledge workers. The first of these models includes high PLA, high intrinsic motivation, single, white, and younger females with no children at home. The second of these two models includes low PLA, high intrinsic motivation, high income, and younger white males that are married with children at home. The main point here is that modeling such complex configurations helps to clarify the low CEP and high CEW knowledge workers.

4.3 *Findings for Models for Low CEW and Low CEP*

Table 4.5c includes four complex antecedent models with consistency for both low CEW and low CEP. Model 1 represents low PLA, low intrinsic motivation, low income, younger, and white females with no children at home. Model 2 represents

Table 4.5b Testing P8b: Knowledge worker identity, perceived location autonomy, and intrinsic work motivation indicate workers high in CEW AND NOT CEP

	Coverage	Consistency
1. Locauto•motv•~inc•white•~marital•~child_home•~age•~gender	0.07	0.90
2. ~locauto•motv•inc•white•marital•child_home•~age•gender	0.07	0.87

P8b: model—cew_notcep = f(locauto, motv, inc, white, marital, child_home, age, gender); frequency cutoff, 2.00; consistency cutoff, 0.87

Table 4.5c Testing P8c: Knowledge worker identity, perceived location autonomy, and intrinsic work motivation indicate workers high in NEITHER CEW NOR CEP

	Coverage	Consistency
1. ~locauto•~motv•~inc•white•~child_home•~age•~gender	0.11	0.90
2. ~locauto•~motv•inc•white•marital•child_home•~age•~gender	0.11	0.89
3. Locauto•~motv•~inc•white•~marital•~child_home•age•gender	0.05	0.86
4. ~locauto•motv•~inc•white•marital•child_home•~age•~gender	0.06	0.88

Model: not(cep_cew) = f(locauto, motv, inc, white, marital, child_home, age, gender); frequency cutoff, 2.00; consistency cutoff, 0.86

low location autonomy, low intrinsic motivation, high income, and married young white females with children at home. Model 3 represents high location autonomy, low intrinsic motivation, low income, older, single, and white males without children at home. In addition, model 4 represents low PLA, high intrinsic motivation, younger, married, low income, and females with children at home.

5 Discussion, Limitations, and Contributions

The study here contributes to moving beyond the issues of whether or not intrinsic work motivation mediates the relationship between PLA and choices of work environment to enhance personal well-being and/or productivity, and the size of the effect of this relationship, and whether or not the impact is always positive. The study contributes to theory and research by proposing that the impact of PLA through intrinsic work motivation on choices of work environment to enhance personal well-being and/or productivity depends on configurations of knowledge worker identity factors. Rather than looking narrowly at PLA, intrinsic work motivation, and choices of work environment to enhance personal well-being and/or productivity, a more complex in-depth stance is necessary. Such a stance makes it possible to learn the combinations of conditions whereby high PLA associates with very high as well as very low intrinsic work motivation and whereby each of these is associated with high and low choices of work environment to enhance personal well-being and productivity. Separately, theory and empirical reports need to consider the configurations associating with very low PLA and very low intrinsic work motivation and choices of work environment to enhance personal well-being and productivity.

Both the first and second models of configurations of knowledge workers that indicated choices high in both CEW and CEP include high PLA, high intrinsic motivation, low income, and older individuals with no children at home (Table 4.5a). This is not surprising, given that the individuals with this configuration feel more free to make choices about where to work and are less impacted by children at home—a condition that may restrict the ability to take advantage of a variety of worksites, including home, to best meet both their personal well-being and productivity goals. The third model includes children at home but also a spouse. In this third model, the knowledge worker is also of the male gender with high income—conditions that seem to offset the impact of low PLA regarding outcomes of high CEW and CEP. Perhaps older generation married couples default to traditional gender roles, which then affords the male partner the opportunity to work outside of the home in a way that still leads to benefits to both personal well-being and productivity through worksite choices. This is consistent with prior research that suggests gendered differences in marital power and resources (Tichenor 1999).

In reviewing the configurations that lead to choices low in both CEW and CEP, it is noticeable how much influence low intrinsic work motivation plays—three of the four models have low intrinsic work motivation. The first model not only has low intrinsic work motivation but also has low perceived freedom to choose places to work (low PLA). These conditions coupled with also being low income lead to few individual resources to make choices that either benefit the individual's personal well-being or their work productivity. Similarly, the second model also has low perceived freedom to choose places to work and although high income, are married and have children at home. Perhaps the combination of limited perceived freedom of choice coupled with the demands from the non-work domain leads to extra knowledge worker strain and the inability to make worksite choices that benefit her personal well-being or productivity. In contrast, the single white males without children at home in model 3 feel free to make choices regarding where to work, but something about being older and not having high intrinsic motivation is enough to limit their ability to find work environments that benefit their personal well-being or their productivity. Perhaps as older male workers, they have internalized social pressure to work in the traditional designated office, even if it offers no performance or well-being benefits (e.g., Stirpe et al. 2018). And finally, in the fourth model, there are very intrinsically motivated young female knowledge workers, but they are married, have children at home, are low income, and do not feel free to choose work environments. Perhaps because of their non-work domain demands and having fewer resources tied to having low income, they perceive greater constraints to their work environment choices, which then limits their ability to seek environments that benefit their personal well-being and productivity.

Frontline, middle, and senior managers may seek to focus on all the complex antecedent conditions associated with work environment choices of knowledge workers—whether those choices benefit the worker's personal well-being, their work productivity, both, or neither. These complex conditions are likely to include recipes that lead to very successful, unsuccessful, or mediocre utilization of flexible work arrangements. Identifying these alternative complex configurations is possible and a worthy objective of theory, research, and practice. Clearly, the present work highlights the variance and complexities in how knowledge workers interpret and implement policies regarding flexible work arrangements. As choosing work environments offers an opportunity for enhancing autonomy, which should result in enhanced intrinsic work motivation and higher levels of productivity and well-being, attention to where the uneven distribution of these benefits lies is an important endeavor. Research to advance theory in this area needs to embrace moving beyond a net-effect symmetric perspective to the construction of useful algorithms for predicting choices of work environments to enhance productivity and well-being as well as for predicting the conditions when choices of work environments diminish productivity and well-being. Such nuanced understanding through the case-based perspective can offer insights into overcoming the blockages in some configurations while highlighting conditions that enhance successful work environment choice outcomes (i.e., beneficial to both productivity and worker well-being) that could perhaps be recreated for those in disadvantaged configurations.

5.1 Confirming the Four Tenets of Configural Theory

The findings support the four tenets of configural theory. The multiple configurations associating (leading to) high CEW and CEP confirm the tenet of equifinality. Consequently, a single configuration may be sufficient, but it is not necessary for accurately predicting high CEW or CEP. The equifinality tenet extends to the valences of simple antecedent conditions; that is, both positive and negative associations for the same simple antecedent condition that appear in different complex antecedent conditions can associate with high scores for cases in an outcome condition. The findings appearing in Tables 4.5a and 4.5b support this perspective. For example, in Table 4.5a, the findings for model 3 with a low PLA and high intrinsic work motivation configuration include outcomes high in both CEW and CEP, but model 4 in Table 4.5b with those same components in the configuration includes outcomes low in both CEW and CEP. The findings confirm the second tenet of configural theory—configural relationships that are accurate in predicting very high CEW and CEP are asymmetric and indicate sufficiency but not necessity. Low scores on these configurations associate with both low and high scores on the two outcome conditions (very low and very high CEP or CEW). This perspective is another way of saying that linear causality is rare in real life, and researchers should discard the view of single-model sufficiency and necessity. Thus, researchers should end reporting in terms of "critical importance" and "key success factors," as no one condition is likely to be both sufficient and necessary.

The findings confirm the third tenet of configural theory. Achieving very high CEW and very high CEP is a complex undertaking. The examination of net effect and two- and three-way interactions is insufficient in describing and understanding very high CEW and CEP. Useful models in describing and understanding such outcome conditions likely require embracing a configural stance such as crafting and testing algorithms via fsQCA.

The findings confirm the fourth tenet. The findings support a causal asymmetry perspective—the configurations associating very high CEP are not the mirror opposites of the configurations associating with very low CEP—the same conclusion applies for CEW. Having a deep accurate understanding of what brings about high CEP only provides partially accurate hints of the causes of very low CEP. Modeling both very high and very low CEP and CEW is necessary.

5.2 Limitations and Suggestions for Future Research

The study here develops and explores configural theory of knowledge workers' choices of work environments to enhance personal well-being and productivity. A current study limitation comes in the form of ethnicity representation in the sample. In the current data set, the sample was mostly comprised of white/Caucasian (88%) knowledge workers. Having only "white" in the models may be a function of having sufficient "white" cases to meet the requirement for high consistency (>0.82) as well as potential variability among the African American/black individuals. Further examination is needed.

Additionally, a general confirmation of the theory and findings for alternative configurations requires going beyond one empirical study in one industry in one country. One helpful step is to replicate the survey both within and beyond the education industry. While the university context is beneficial in providing an organizational setting devoid of the often more intensive privacy and information security concerns found in other industries, these other industries may provide additional insight into factors and configurations that affect CEP and CEW among knowledge workers. Additionally, the current sample is US-based, so this research could strongly benefit from replication in other country contexts.

The findings of the current study suggest that myriad ways occur in which knowledge worker identity factors influence the perception and utilization of flexible work arrangements that afford choice regarding work environments. Research is necessary to identify the mechanisms by which these factors influence the relationships tested here, in order to responsively adapt policy and practice to enhance individual success in navigating work environment choices.

5.3 Implications

The current study demonstrates that many underlying reasons explain why organizations may not achieve universally successful flexible work programs, such as those that offer choice regarding where to work. Specifically, this study highlights the individual way in which policies and practices are navigated based on configurations of individual characteristics, rather than simplistic presence/absence of variables revealed by symmetric studies. Taken together, these findings offer insight into previously discrepant findings and industry experiences regarding the overall failures and successes of offering worksite flexibility.

As an example, the findings of the current study challenge the results of previous works that found no relation between worker age and benefits of flexible work arrangements. Specifically, while Stirpe et al. (2018) reported a nonsignificant association between flexible work arrangements and employee retention in more aging workplaces, this study offers a counterargument. As the results presented here show, it is likely that there are unique configurations of age plus other individual level factors, such as their interpretation of the flexible work arrangements, which indicate higher (lower) levels of employee retention. Without using a complexity theory lens, these configurational models are obscured and can lead to misleading conclusions regarding age as a meaningful identity factor. These previous studies should be reexamined using a similar approach to seek such configurations.

However, great care is necessary in translating the findings of this research into actionable insights by managers and organization leaders. It is important to note that complexity theory, configural analysis, and the findings in this article should not be used to exclude knowledge workers from access to location autonomy in their jobs because their identity profiles associate with low CEW and CEP. The moral issues and unintended negative consequences of such decisions can overwhelm any immediate benefits from applying such profiling algorithms (cf. Carter 2003). The moral and long view that deserves management attention may be that organizations accept a substantial measure of responsibility in investing in facilitating the intended consequences of offering flexible work arrangements—that is, recognizing and positively influencing how those arrangements are perceived and enacted by unique individuals within the organization.

References

Carter, P. L. (2003). 'Black' cultural capital, status positioning, and schooling conflicts for low-income African American youth. *Social Problems, 50*(1), 136–155.

De Haaff, B. (2017, October 30). Remote workers are outperforming office workers–here's why: Research shows that office workers cannot concentrate at their desks. Inc. Retrieved from https://www.inc.com/brian-de-haaff/3-ways-remote-workers-outperform-office-workers.html

Deci, E. L., & Ryan, R. M. (1985). *Intrinsic motivation and self-determination in human behavior*. New York: Plenum Press.

Fell, S. F. (2018, January 31). Remote workers are more loyal: 3 insights why these top 100 companies hire remote workers. Inc. Retrieved from https://www.inc.com/sara-sutton-fell/3-reasons-why-these-top-100-companies-hire-remote-workers-why-you-should-too.html

Feurer, S., Baumbach, E., & Woodside, A. G. (2016). Applying configurational theory to build a typology of ethnocentric consumers. *International Marketing Review, 33*, 351–375.

Grzywacz, J. G., & Marks, N. F. (2000). Reconceptualizing the work-family interface: An ecological perspective on the correlates of positive and negative spillover between work and family. *Journal of Occupational Health Psychology, 5*(1), 111–126.

Guest, D. E. (2017). Human resource management and employee well-being: Towards a new analytic framework. *Human Resource Management Journal, 27*(1), 22–38.

Hsiao, P. H., Jaw, C., Huan, T. C., & Woodside, A. (2015). Applying complexity theory to solve hospitality contrarian case conundrums: Illuminating happy-low and unhappy-high performing front line service employees. *International Journal of Contemporary Hospitality Management, 27*(4), 608–647.

James, G. (2018, August 10). Case closed: Work-from-home is the world's smartest management strategy: Research from Stanford and other sources reveals that working from home vastly increases productivity. Inc. Retrieved from https://www.inc.com/geoffrey-james/case-closed-work-from-home-is-worlds-smartest-management-strategy.html

Loretto, W., & Vickerstaff, S. (2015). Gender, age and flexible working in later life. *Work, Employment and Society, 29*, 233–249.

McClelland, D. C. (1998). Identifying competencies with behavioral-event interviews. *Psychological Science, 9*, 331–339.

Misangyi, V. F., Greckhamer, T., Furnari, S., Fiss, P. C., Crilly, D., & Aguilera, R. (2017). Embracing causal complexity: The emergence of a neo-configurational perspective. *Journal of Management, 20*, 1–28.

Mol, C. V. (2017). Improving web survey efficiency: The impact of an extra reminder and reminder content on web survey response. *International Journal of Social Research Methodology, 20*, 317–327.

Nagy, G., Megehee, C. M., Woodside, A. G., Laukkanen, T., Hirvonen, S., & Reijonen, H. (2017). Achieving requisite variety in modeling firms' strategy heterogeneities: Explaining paradoxical firm-market performances. *Industrial Marketing Management, 65*, 100–128.

Ragin, C. C. (1999). Using qualitative comparative analysis to study causal complexity. *Health Services Research, 34*(5), 1225–1239.

Ragin, C. C. (2008). *Redesigning social inquiry: Fuzzy sets and beyond*. Chicago: Chicago University Press.

Spivack, A. J., & Milosevic, I. (2018). Perceived location autonomy and work environment choice: The mediating influence of intrinsic motivation. *Journal of Applied Behavioral Science, 54*(3), 325–348.

Stirpe, L., Trullen, J., & Bonache, J. (2018). Retaining an ageing workforce: The effects of high-performance work systems and flexible work programmes. *Human Resource Management Journal, 28*(4), 585–604.

Tichenor, V. J. (1999). Status and income as gendered resources: The case of marital power. *Journal of Marriage and Family, 61*(3), 638–650.

Tremblay, M. A., Blanchard, C. M., Taylor, S., Pelletier, L. G., & Villeneuve, M. (2009). Work extrinsic and intrinsic motivation scale: Its value for organizational psychology research. *Canadian Journal of Behavioral Science, 41*, 213–226.

Truss, C., Conway, E., d'Amato, A., Kelly, G., Monks, K., Hannon, E., & Flood, P. C. (2012). Knowledge work: Gender-blind or gender-biased? *Work, Employment and Society, 26*, 735–754.

Urry, J. (2005). The complexity turn. *Theory, Culture & Society, 22*, 1–14.

Williamson, I. O., & Cable, D. M. (2003). Predicting early career research productivity: The case of management faculty. *Journal of Organizational Behavior, 24*, 25–44.

Woodside, A. G. (2014). Embrace·perform·model: Complexity theory, contrarian case analysis, and multiple realities. *Journal of Business Research, 67*, 2495–2503.

Zeijen, M. E. L., Peeters, M. C. W., & Hakanen, J. J. (2018). Workaholism versus work engagement and job crafting: What is the role of self-management strategies? *Human Resource Management Journal, 28*(2), 357–373.

Chapter 5
Customers' Assessments of Retail Traditional Local Markets: Strategy Outcome Performance Screening

Jaesuk Jung, Eunju Ko, and Arch G. Woodside

Abstract This study assesses customers' assessments of place sustainability and place attractiveness of a large-scale traditional local market (TLM). The study proposes a general theory of antecedents and outcomes for customer assessments of place sustainability and place attractiveness for one category of TLM: a collection of many independent shops (e.g., 1000 to 10,000+ shops) available contiguously in one geographic area (i.e., a long-period market)—spanning generations of shop-keeping family stores, large in scale and scope, and selling a wide range of products of very substantial varieties, with many shops marketing to ultimate consumers, retailers, or a mix of retailers and consumers. Rather than using the currently dominant theory and research logic of proposing and testing symmetrical directional hypothesis (i.e., increasing X associates with increases in Y) and the reporting of net effects of independent variables, the study reports generalizable asymmetric case-based models of TLM sustainability and attractiveness. Case-based modeling proposes that certain screening rules of antecedent conditions enable identifying cases having certain outcomes consistently (e.g., frequent visits and high expenditure outcomes). The study includes findings from a survey of customers ($n = 162$) of their assessments of place sustainability and place attractiveness of Seomun ("West Gate") Market, a large TLM in Daegu City, Republic of Korea. The findings support the general conclusions that TLM place sustainability and place attractiveness models associate with shoppers having high rates of visiting and expenditures consistently as well as additional models indicating that certain place sustainability and attractiveness models associate with cases having infrequent visits and low expenditures.

This work was supported by Global Research Network program through the Ministry of Education of the Republic of Korea and the National Research Foundation of Korea (NRF-2017S1A2A2041810).

J. Jung (✉) · E. Ko (✉)
Yonsei University, College of Human Ecology, Seoul, Republic of Korea
e-mail: ejko@yonsei.ac.kr

A. G. Woodside
Yonsei University, Yonsei Frontier Lab, Seoul, Republic of Korea
e-mail: arch.woodside@bc.edu

A. G. Woodside (ed.), *Accurate Case Outcome Modeling*,
https://doi.org/10.1007/978-3-030-26818-3_5

Keywords Case · Local markets · Traditional · Place attractiveness · Place sustainability · Shoppers

1 Introduction

This study constructs theory for examining shoppers' and merchants' assessments of place sustainability and place attractiveness of iconic regional traditional local markets (TLMs). Both place sustainability and place attractiveness are multidimensional concepts, and this study proposes multidimensional perspectives for both concepts. A general definition of sustainability is the property of biological systems to remain diverse and productive indefinitely. For TLMs, "place sustainability" is a multidimensional concept including maintenance and development of economic, social, environmental, and cultural well-being of a configuration actors, processes, and outcomes at a site. "Site" here refers to a definable geographic area, though sites may represent a virtual place. "Place attractiveness" here refers to actors' assessments of the desirability of visiting a TLM. Place attractiveness is a multidimensional concept in the present study and incudes place image, satisfaction, satisfaction in visiting and shopping in the TLM, trust, and loyalty. "Actor" encompasses three categories: shoppers, visitors or tourists, and merchants with shops in the TLM.

The study contributes to the retailing literature by proposing and testing a general theory of shoppers' and merchants' assessments of both place sustainability and place attractiveness for a TLM. This study examines the antecedents and outcomes of place sustainability and place (as a distinct place brand) attractiveness assessments by customers and merchants in the context of traditional local markets (TLMs). The study contributes a case-based theory of antecedents and outcomes for both sustainability and place attractiveness actor assessments. "Visitors" or "tourists" refer to actors entering a TLM infrequently; most TLM visitors have no plans to shop in the TLM; most visitors have low-to-zero expenditures during their visit in the TLM. In this study, "merchant" encompasses TLM shop owners in three business categories: business-to-business, business-to-consumers, and a mix of the first two categories with neither category representing more than 75% of shop's revenues.

While recognizing many nations have small-scale TLMs that include less than 100 shops, the study here applies the following definition to TLMs: A TLM is a collection of many independent shops (e.g., 1000 to 10,000+ shops) available contiguously in one geographic area, which are long-period markets—spanning generations of shop-keeping family stores, large in scale and scope, and selling a wide range of products of very substantial varieties, with many shops marketing to ultimate consumers, retailers, or a mix of retailers and consumers. "Grand Bazaar" in Istanbul represents an iconic TLM. The Grand Bazaar is one of the largest and oldest covered markets in the world, with 61 covered streets and over 4000 shops which attract between 250,000 and 400,000 visitors and shoppers daily. In 2014, the Grand Bazaar was first among world's most-visited tourist attractions with 91 million annual visitors (Appleton et al. 2014). The present study focuses mostly on examining the antecedents and outcomes of assessments of one TLM, Seomun ("West

Gate") Market. Seomun Market is located in the center of Daegu City, the third largest city in the Republic of Korea (i.e., South Korea). Seomun Market includes more than 4000 shops in 6 main buildings and outdoor street sites. The present study focuses on examining antecedents and outcomes of assessments by shoppers. One shop owner (Jung 2017) estimates less than 10% of persons entering Seomun Market annually are tourists; this estimate was the modal response given by five additional shop owners interviewed in the study.

The present study addresses the following issues: Are shoppers identifiable by demographic configurations for both their global assessments of TLM sustainability and by each of the four dimensions of TLM sustainability? Are shoppers identifiable by demographic configurations for both their global assessments of TLM attractiveness and by each of the four dimensions of TLM attractiveness? Shopper identification refers to specific demographic screens. The study focuses on constructing and testing asymmetric case-based models of antecedent configurations identifying high (e.g., cases in the top quintile) and low (e.g., cases in the bottom quintile) of sustainability. As described below, the study applies complexity theory tenets including the tenet that the causes of high outcome membership scores differ in configurations from the causes indicating low outcome membership scores.

The study addresses the following additional issues: Are case-based demographic models accurate in identifying shoppers with high monthly visits to the TLM? Are global as well as dimensional place sustainability and place attractiveness models accurate in identifying shoppers with high monthly expenditures in the TLM? Does the combination of demographic dimensions with sustainability dimensions identify shoppers consistently having high numbers of visits? Does the combination of demographic dimensions with sustainability dimensions identify shoppers consistently having high expenditures? Similar questions apply for negative outcome cases, for example, can models be constructed that identify cases having low scores in visits consistently?

This study moves beyond the use of null hypothesis statistical tests (NHSTs) of directional symmetric relationships (e.g., more versus less positive assessments of sustainability associate with more shopping trips to a TLM). While still the dominant logic in theory construction and testing via regression analysis (e.g., stepwise multiple regression and structural equation models), thorough examinations of the usefulness of NHST indicate their use in theory construction and data analysis represents bad science practice (cf. Hubbard 2016; Trafimow 2014; Woodside 2016a, b, Woodside 2017; Ziliak and McCloskey 2008). Statistical significance is achievable for any difference from the null simply by increasing the sample size (e.g., from $n = 300$ to $n = 3000$). Studies using symmetric tests (e.g., correlations, F-tests, MRA (multiple regression analysis), and SEM (structural equation modeling)) almost always do not consider case anomalies, that is, cases having the opposite relationship from the significant main effect reported. The current pervasive practice of theory construction to identify cases having certain properties and testing for statistical significance of relationships among variables is bad due to its mismatch of statements regarding cases and tests regarding variables (Fiss 2007, 2011). Given data analyses support complexity theory tenets consistently (discussed below), NHST represents a family of shallow theory construction and data analytic tools;

researchers can leave the shallows by embracing somewhat precise outcome testing (SPOT) that identifies cases using "computing with words" (CWW, Zadeh 1996, 2010) models (i.e., screens that identify cases having certain properties that indicate a specific outcome consistently). The present study applies CWW and SPOT in asymmetric theory construction and testing of shoppers' TLM place sustainability and attractiveness antecedents and outcomes.

Case-based modeling consists of building one or more screens that accurately identify cases that meet or surpass all conditional statements appearing in the screens. For example, older (age), married (marital), high-income (income) females (~Gender) who are very satisfied (satisfy) with a TLM visit ("visit") the TLM frequently. Model 1 expresses this hypothetical model:

$$\text{age} \bullet \text{marital} \bullet \text{income} \bullet {\sim}\,\text{gender} \bullet \text{satisfy} \;''\; \text{Visit} \tag{5.1}$$

The model appears in the left side of Eq. 5.1. The mid-level dot ("•") represents the logical AND operation in Boolean algebra. The sideways tilde ("~") indicates negation; if male is calibrated to equal 1.00, then the tilde indicates female. Applying a calibration procedure whereby all raw values for variables are calibrated (i.e., transformed) using a logistic function, the values become membership scores in a condition ranging for 0.00 to 1.00. A membership score for age equal to 1.00 indicates old age member. Equation 5.1 is a fuzzy statement; fuzzy statements are "computing with words (CWW)" (Zadeh 1996, 2010). The logistic function calibrates raw values into membership scores for each condition whereby the membership scores have a lazy "s" ("$\int$") distribution. A case's membership score for a model is equal to the lowest calibration score among the conditions in the model. Hypothetical examples: Mary is 68 years old, married, in a household in the top quintile income level, very satisfied with the TLM and visits the TLM twice a week (visit rate is calibrated into an outcome score); Judy is 78 years old, married, in a household with median level income, somewhat dissatisfied with the TLM and visits the TLM once a month. Here are Mary's and Judy's hypothetical calibrated scores for each antecedent condition and outcome condition for model 1.

Case	Age	Marital	Income	~Gender	Satisfied	Model score	Outcome score
Mary	1.00	1.00	1.00	1.00	1.00	1.00	1.00
Judy	1.00	1.00	0.50	1.00	0.10	0.10	0.25

The model score for each case is equal to the lowest membership score among the conditions in the model. Thus, Judy's model score is equal to her satisfaction score because her satisfaction score is the lowest score among the five conditions in the model. Thus, the logical "AND" condition represents how much membership a case shares among the conditions in a model. For age, using a tipping point equal to 65 years for full membership in old age, both Mary and Judy have the same calibrated age (1.00). The accuracy of the model is consistent for both cases, that is, the model scores for Mary and Judy are equal to or less than the outcome score of visit frequency. Writing and testing the accuracy of fuzzy statements is an example of computing and testing with words (CWW). Examining the accuracy of a fuzzy statement in predicting scores for samples of new cases from the cases used first to examine the accuracies of alternative models is an example of SPOT (Woodside 2016b, 2017). Case-based model builders typically use a high bar of predictive accuracy (e.g., a 4-to-1 accurate identification of cases high in the outcome condition) in deciding that a model is useful. The method section below and Ragin (2008) include additional details on how to calibrate values into membership scores and how to estimate the accuracy of models.

Following this introduction, Sect. 2 presents the general theory of place sustainability and place attractiveness and includes a brief review of literature relating to place (e.g., shopping mall) assessments by customers. Section 3 is a discussion of complexity theory tenets that are relevant to case-based theory construction and testing in the present study and in asymmetric modeling in general. Section 4 presents the method for collecting assessment and related data on shoppers and merchants in the Seomun Market. Section 5 presents and discusses the findings. Section 6 discusses the contribution to theory and TLM management practice and includes limitations and suggestions for additional research. Section 7 presents conclusions.

The Republic of Korea ("South Korea") is a developed, high-tech nation, where its retailing industry includes distinct "cultural strategy" ideologies with features in their configurations differing profoundly from one another: modern mega-Western-style malls as well as local traditional markets (TLMs). In contrast to Seomun representing a Korean TLM, Starfield Hanam is a mega-Western-style mall or "shopping theme park" ("Imagine your Korea" 2017) with a 456,000-square-meter (4.9 million-square-foot) retail location with a gross leasable area of 158,000 square meters (1.7 million square feet). It opened in September 2016 as South Korea's largest Western-style shopping mall. Starfield Hanam houses some 800 store brands as well as hypermarkets, dining establishments, and a wealth of cinema, sports, and other entertainment facilities. Korean Western-style, retail sites are typically vertical and dominated by department stores. During the inauguration year of Starfield Hanam, the developers expected 12 million visits annually and achieved this outcome.

Possibly, a share (10% to 30%) of shoppers in TLM and mega-Western-style retail complexes may be the same persons, but shoppers segmented by "share of wallet" would likely indicate that most customers who are frequent shoppers are one, but not both, of these distinctly different retailing subcultures. The objectives of the present study do not include empirical comparisons of visitors' and shoppers' place sustainability and attractiveness assessments for each of these retailing performing subcultures. The present study tests the perspective that actors having specific demographic antecedent configurations indicate place sustainability configurations with high memberships in TLM sustainability cultures versus actors in alternative specific demographic configurations indicating place sustainability configurations with high memberships in mega-Western-style (MWS) retailing sustainability culture. Theory development here includes the perspective that both TLM and MWS have distinguishable sustainability subcultures. Table 5.1 includes the configurational features in the TLM versus MWS subcultures and the perspective place sustainability configurations that each subculture supports.

Table 5.1 Configurations of retailing subcultural strategy and dimensions of place sustainability assessment

Retailing subculture ideology		
Retailing subculture condition	MWS (mega-Western-style) mall	(TLM) traditional local market
1. Design	Symmetrical	Asymmetrical
2. Focus	Brand, luxury, fame	Original, sophistication, exotic
3. Availability	Ubiquitous	Rare, requires seeking out
4. Technology	Modern scientific	Pre-modern traditional
5. Production	Mass scale	Small lots, special batches
6. Service experience speed	Fast	Slow
7. Psychological representations	Narcissism, elite, celebrity	Discernment, experience, insight
8. Drama enactment theme	Pampering qualified shoppers	Co-creating authentic experience
9. Principal sustainability orientation	Economic and not environmental	Environment, social, and not economic
10. Cultural code	Psycho-analytical, archetypes (surface)	Pre-industrial cultural detail configuration (depth, substance)
11. Exemplar	Versace stores, fashion, clothing	Darye (traditional Korean tea ceremony)

Source: Original to the present study

Both and TLM and MWS represent configurations of tangible and nontangible features that enable highly distinct shopping experiences and outcomes. For some customer shopping segments (not all shoppers) may indicate high sustainability assessments, but these assessments are likely to differ from one another to an extraordinary extent (row 9 in Table 5.1). The present study examines and supports this proposal for TLM shoppers. The examination supports the conclusion that high shopper sustainability assessments occur for unique sustainability configurations and not for all subdimensions of place sustainability.

2 A Case-Based General Theory of Place Sustainability and Place Attractiveness

Theory construction in this study builds from Gigerenzer's (Gigerenzer 1991: 19) astute observation: "Scientists tools are not neutral." That is, data analytic tools impact the issues researchers seek to answer in their theory construction. The dominant logic today (2018) is researchers in the sub-disciplines of business, psychology, and sociology construct theories that they plan to examine via symmetric tests (i.e., NHSTs); their objective is to show that hypothesized relationships are statistically significant from zero and to report and compare net effect sizes of relationships.

Researchers are using case-based theory construction and testing when proposing screens that should be accurate consistently (according to their theory) in identifying cases exhibiting certain outcomes (e.g., highly satisfied customers (Wu et al. 2014), highly competent managers (McClelland 1998), and adopters of a new hotel service (Ordanini et al. 2014)). "Consistently" refers to constructing models identifying cases all (or nearly all) having the same outcome. Such case-based researchers constructing screens as a tool embrace Hubbard's (Hubbard 2016: 192) stance: "In principle, there is no reason why theories in the management and social sciences cannot yield precise [or somewhat precise] (or interval) predictions."

Modeling precise predictions calls for forecasting case outcomes rather than directional relationships among variables. Modeling to identify cases representing an outcome of interest (e.g., cases in the top quintile for expenditures) involves discretizing values of all variables into conditional membership scores. Membership scoring by quintiles (e.g., McClelland 1998) or by 100-point calibration (a transformation of values into membership scores) is a necessary step in such case-based forecasting models. Case-based modeling offers hints to researchers to embrace a theoretical paradigm radically different from the now pervasive symmetric model construction and testing (Misangyi et al. 2017; Woodside 2014).

Core complexity theory tenets are foundational for the radically innovative case-based paradigm. Section 3 is a brief primer of core complexity theory tenets. "When recipes are more important than the ingredients" (Ordanini et al. 2014) follows from adopting complexity theory tenets and examining data by membership scores and screening algorithms. With reasonably large sample sizes (e.g., $n > 100$), following transforming continuous variables into quintile scores and cross-tabulating each independent (antecedent = X) score with a dependent (outcome = Y) score, cases will be observable usually in all 25 cells of all cross-tabs even when the main effect for XY is large. Consequently, case-based researchers attempt to model each of the four corners: when high X indicates high Y, when high X indicates low Y, when low X indicates high Y, and when low X indicates low Y. For all X and Y comparisons, their relationships are almost never entirely symmetrical. Rather than viewing cases indicating a negative relationship between X and Y as unexplainable blips when the main effect for X and Y is large (e.g., $r^2 > 0.50$), case-based researchers raise and attempt to answer the questions of when each of the four-corner case outcomes occurs.

Figure 5.1 is a visual synopsis of a case-based general theory of place sustainability and place attractiveness. The Venn diagrams in Fig. 5.1 illustrate possible configurations of simple and complex antecedent and outcome conditions. Figure 5.1 includes four sets of conditions that include demographics, place sustainability, place attractiveness, and visit and expenditure outcomes. Arrows in and out of place sustainability and place attractiveness in Figure indicate that both are complex antecedent as well as outcome conditions. The arrows in Fig. 5.1 represent theoretical propositions of the theory. The objective of the general theory is to explain and predict intermediate and final cases in the outcome conditions. Identifying cases high in a simple or complex outcome condition is the objective of case-based modeling. Cases (e.g., shoppers) very high (e.g., top quintile) in satisfaction with a TLM

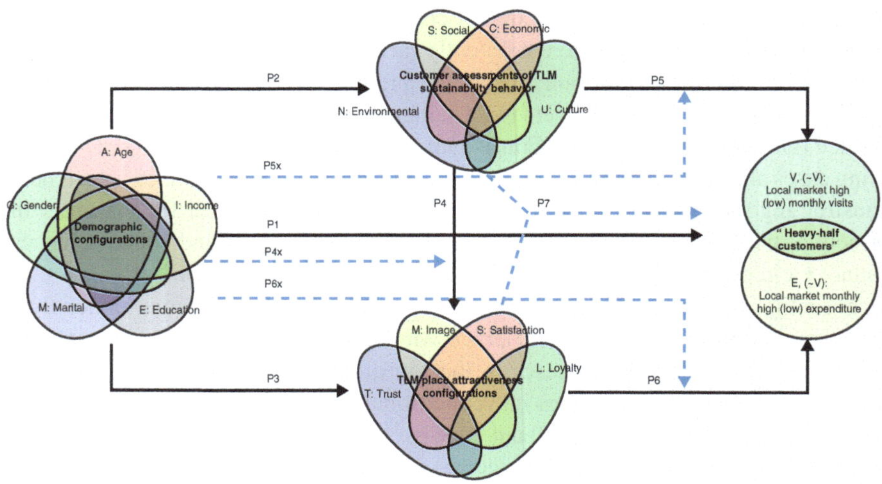

Fig. 5.1 Customer (actor-based) assessments of traditional local market (TLM) sustainability behavior and TLK place attractiveness: theory. Key: "~" is the negation of the condition. Note: Theory includes asymmetric models; thus, both V and ~V are separate model outcomesThe theory models E and ~E separately as well

is an example of a simple outcome. Cases very high in the negation of satisfaction (e.g., bottom quintile) are another simple outcome condition. Substantial empirical evidence supports the general theory's proposition that relationships are asymmetric rather than symmetric even when the effect size of an XY relationship is large: Often cases low in X do not have low scores in Y when a causal condition showing most cases high in X are also high in an outcome Y. The inclusion of ingredients in causal configurations indicating cases with high Y outcomes differ from the ingredients (not just direction) from the causal configurations indicating cases having low membership in Y (i.e., ~Y). Thus, the study of how firms succeed provides scant knowledge about how firms fail. The study of causal antecedent configurations for shopping in a TLM provides scant knowledge about the causal antecedent conditions for not shopping in a TLM.

This general theory of place sustainability and place attractiveness assessments builds from a foundation of major tenets of complexity theory. Complexity theory tenets (C_i) are supportable by empirical observations. For example, cases very high in a simple antecedent condition are an insufficient indicator that these cases are very high in an outcome condition consistently. Stating a configuration of necessary and/or sufficient combination of conditions (i.e., ingredients or features) explicitly or implicitly is sufficient for indicating a somewhat precise outcome consistently. A configuration indicates membership is two or more simple conditions. For example, in Fig. 5.1, a young male having high income who did not graduate from high school is a configuration of three simple conditions.

Performing a thought experiment is useful for illustrating the use of case-based modeling. Thus, consider the possible combinations of five demographic antecedent conditions and one outcome condition (e.g., visit frequency per month to a TLM). The outcome (e.g., visits per month or expenditures per month) is discretized by quintiles. The present study identifies 768 logical configurations of 5 demographic conditions: 3 by 4 by 4 by 4 by 4 = 768 demographic configurations and 3840 total configurations taking into account 5 levels of an outcome condition—discretized visit frequency into 5 quintiles. The logical configurations include three possible gender conditions: married, single, and irrelevant in a complex condition indicating cases with high scores in an outcome. Frequently, in case-based modeling, one or more accurate models occur that applies to the top quintile in the outcome condition, and other case-based models are relevant for case-based models of cases in the bottom quintile. For age, income, education, and marital status, the study has four conditions: high (top 20%), ambiguous, low (bottom 20%), and absent. For example, for marital status, a person not married has a low membership (0.00), a person separated from a spouse has an ambiguous membership (0.50), a married person has a high marital score (1.00), and the absent condition refers to marital status being irrelevant in a model—marital status is absent from a model that accurately indicates a high outcome consistently. A complex configuration with high "consistency" is a model with all or nearly all cases having top (bottom) quintile membership. For example, assume that 12 cases are available for testing model 1 (age•marital•income•~gender•satisfy ≤ visits) and 11 of that 12 cases are in the top quintile for visiting a TLM. The model indicates that more than 9-to-1 shoppers who are members of this complex condition are very frequent visitors to the TLM. The screen indicates a given outcome with high consistency. Assuming that 50 cases are very high (top quintile) in visits, a model that identifies 11 cases correctly as being very high in visits has a coverage equal to 0.22 (i.e., 11/50). (Increasing the complexity in case-based modeling by increasing the number of ingredients in a configuration serves to decrease coverage but increase consistency.)

Data are clumpy with many cases located in a limited number of cells, and many cells occur with zero or only one or two cases. Even though a sample size is very large (e.g., n = 1000), a cross-tabulations of five conditions resulting in 768 cells will frequently include 50 to 100 plus empty cells and some cells with 10 to 50 cases. A "truth table" is the identification given to all logically possible combinations of simple conditions including the empty cells and the cells with one or more cases. Frequently, cells with a limited number of cases (i.e., low coverage) are a special interest to researchers. Cases in cells indicating reversal outcomes to a main effect often represent anomalies. "An anomaly is a fact that doesn't fit received wisdom…an anomaly marks an opportunity to learn something very valuable. In science, anomalies are the frontier, where the action is" (Rumelt 2011, pp. 247–248). Researchers performing symmetric tests only do not recognize the occurrence of anomalies. Based on our experience in performing case-based modeling, researchers should expect to find anomalies between 6% and 15% of cases in a study even when a relationship between variables indicates a large effect size. Two types of anomalies for an XY relationship occur: false-negative and false-positive cases. For example, assuming a high positive relationship occurs for income and visits to a TLM, a false negative is a shopper with low income and frequent visits. A false positive is a shopper with high income and infrequent visits. A "four-corner" analysis (Woodside et al. 2018) is constructing antecedent models for each of the four outcome configurations: high income and high visits, low income and low visits, high income and low visits, and low income and low visits.

Figure 5.1 includes seven main propositions that describe a few complex configurations that identify shoppers with frequent visits and another set of complex configurations that identify shoppers with high expenditures. This approach to modeling outcomes is also capable of identifying negation outcomes—shoppers with infrequent visits and shoppers with low expenditures. Four-corner analysis is possible as well: identifying shoppers who visit infrequently with high expenditures, shoppers who visit every week with low expenditures, frequent shoppers with high expenditures, and infrequent shoppers with low expenditures. The three additional propositions with an "x" in Fig. 5.1 (P4X, P5X, and P6x) indicate that consistencies for place sustainability and place attractiveness as antecedents should be high for shoppers with specific demographic configurations. Thus, demographic configurations may have an indirect rather than a direct impact on shoppers with frequent (or infrequent) visits and high (or low) expenditures.

Researchers can state complex antecedent conditions in general and specific configurations as the following sub-sections illustrate. Just as variable-directional researchers should avoid using stepwise multiple regression analysis because of the shallow perspective such an analysis implies (Armstrong 2012), case-based researchers should avoid relying solely on software subroutines to examine all logically possible configurations. The practice pervasive right now is not to test for all possible configurations to find the few that indicate top quintile or similar type cases rather than proposing any specific configurations. Based on complexity theory tenets including the occurrence of surprising reversals from statistically significant main effects and because studies are scarce on case-based modeling for identifying shoppers exhibiting certain outcomes in comparison to variable-directional modeling of net effects, the present study examines proposals for both general and specific configurations.

2.1 P1: Complex Antecedent Demographic Configurations and Visit Frequency and Expenditures

The majority of studies focused on explaining and predicting patronage behavior in retailing include a table showing survey respondent's distributions of demographic characteristics but do not attempt to include demographic characteristics in their patronage forecasting models. Ogle et al. (2004), Wu et al. (2014), and the present study are exceptions to this observation. The general demographic proposition are (P1a) a few complex antecedent configurations of demographic characteristics identify shoppers who are consistently (a) very frequent visitors or (b) have very high expenditures and (P1b) a few complex antecedent configurations of demographic characteristics identify shoppers consistently who are (a) very infrequent visitors or (b) have very low expenditures. Based on the complexity theory tenet that the causal antecedent for success differs from the causal antecedents for failure, distinctly different models are proposed for frequent versus infrequent shopping and high versus low shopping expenditures. Information on shoppers' perspectives regarding place sustainability and/or place attractiveness is necessary to accurately predict these outcomes. The few available early (Stone 1954) and more recent (Wesley et al. 2006) case-based modeling studies on shopping outcomes support this general proposition that certain demographic identities influence retail shopping behavior. However, the modeling approaches used by Stone (1954), Ogle et al. (2004), and Wesley et al. (2006) do not include formal tests of specific case-cased models. Variable-based directional tests of demographic variables (e.g., Ogle et al. 2004) report inconclusive findings of demographic influences on patronage because they do not examine demographics alone either for simple correlations with shopping intentions toward one retail firm or in a multiple regression analysis using only demographic terms; Ogle et al. (2004) examine the effects of 4 demographic variables (age, gender, income, and education) as terms along with psychological variables (e.g., shopping orientation and attitude toward the behavior) in regression models having 15 terms and do not report correlations of any of the variables with shopping intentions. Wesley et al. (2006) are able to identify cases for high expenditures in mall shopping by configurations of demographic and psychological conditions.

The present study includes proposing two specific complex demographic antecedent conditions that indicate frequent shopping trips into the TLM where A is high age, G is male gender (thus, "~G" indicates female), I is high income, E is high education, M is married, and V is shoppers with high visit frequency:

$$A \bullet \sim G \bullet I \bullet E \bullet \sim M \;''\; V \tag{5.2}$$

CWW: Old, unmarried, high-income females, with high education (this segment of cases has the time and inclination to shop), have frequent visits to the TLM.

$$\sim A \bullet \sim G \bullet I \bullet \sim E \bullet \sim M \leq V \tag{5.3}$$

CWW: Young, unmarried females with low education and high income (cases fitting this screen visit the TLM to meet friends) have infrequent visits to the TLM.

The study includes proposing and testing two specific models for infrequent visits to the TLM:

$$A \bullet G \bullet \sim I \bullet E \bullet \sim M \;''\! \sim V \tag{5.4}$$

CWW: Older unmarried males with high education and low income (this segment has alternative pursuits and hobbies (e.g., library and online visiting with friends)) have infrequent visits to the TLM.

$$\sim A \bullet \sim G \bullet \sim I \bullet \sim E \bullet M \leq \sim V \tag{5.5}$$

CWW: Young married females with low incomes and low education (this segment may need to be home with their husbands and possibly with children) have infrequent visits to the TLM.

2.2 P2: A Few Complex Antecedent Demographic Configurations Affect Place Sustainability

P2a: A few complex demographic configurations indicate customers high in overall, "macro" place sustainability. P2b: A few complex demographic configurations indicate customers high in each of four dimensions of place sustainability. P2c: Cases represented by a few complex demographic configurations indicate the TLM to have low macro-sustainability performances. P2d: Cases represented by a few complex demographic configurations find the TLM to have low micro (each of four dimensions) sustainability performances. "Sustainability" refers to recognizing and performing actions that nurture the well-being of multiple stakeholders participating directly and/or indirectly in interconnecting behaviors while not degrading the well-being of one or more stakeholders and natural environments. "Sustainability science" is the field of research focusing on describing, explaining, predicting, and controlling the well-being and preventing the degradation of stakeholders engaging in interconnected behaviors (cf. Kates et al. 2001). Literature relevant to the present study includes examinations of four principal dimensions of sustainability: social, economic, environmental, and cultural.

Social Sustainability McKenzie (2004) offers a useful though over-restrictive definition of social sustainability in the contexts of TLMs: "Social sustainability occurs when the formal and informal processes, systems, structures, and relationships actively support the capacity of current and future generations to create healthy and livable communities. Socially sustainable communities are equitable, diverse, connected and democratic and provide a good quality of life." TLMs in some nations do not include the features of high diversity and democracy. Behavior and categories of shops, number of shops, and national origins of shops are sometimes highly restrictive by a TLM central commission or family governing board (Park 2012). Because many-to-most customers visit the same TLM and same shops in the TLM on the majority of days each week and customers and shop retailers greet each other warmly and frequently by name, the a priori expectation was the responses to the measurement scale for social sustainability would be positively biased above the scale's midpoint (i.e., 12, for the 4- item, 5-point scale). In the present study, the findings supported this expectation partially (75% of customers provided responses with assessments above the scale's midpoint). In the data set, 6 percent of the customers gave negative (scores 8 or less) assessments on the social sustainability scale. Such a distribution of responses for social sustainability and similar distributions of responses for the other scales support the need to adopt a complexity theory foundation for the study of antecedents and outcomes to sustainability in the context of TLMs.

Economic Sustainability Economic sustainability refers to achieving micro- and macro-economic and financial objectives by the actions by the organization (firm, market) that nurture the organization's long-term survival. Micro objectives include decisions/actions to maintain its (e.g., market's) long-term survival; providing returns adequate to support its stakeholders' objectives including income and health benefits of employees and survival of its suppliers; adopting new technologies necessary for survival; responding effectively to actions by competitors that may negatively impact the organization's financial success; and achieving operating efficiencies capable of attracting resources for maintaining its economic well-being. Macro objectives include contributing to economic well-being of the surrounding communities and indirect stakeholders that affect the survival of its immediate, direct stakeholders—and doing so without damaging the economic survival of its direct and indirect stakeholders.

For TLMs, economic growth is not a necessary condition for economic sustainability; analogous to Weick and Sutcliffe (2007) highly reliable organization (HRO) theory and operational stance versus financial growth objectives, an TLM may implicitly or explicitly identify providing highly functioning organization (HFO) that includes providing livable (not maximum) economic returns as the dominant economic sustainability objective. Consequently, a "profit maximization" objective is not a characteristic identifying high TLM economic sustainability.

Environmental Sustainability Retailing environmental sustainability refers to several behaviors by retailers including buying "green" or naturally manufactured, not synthetic, for resale; not polluting natural environments (e.g., air, water, and soil) in manufacturing and reselling of products and services; and providing and advocating the use of environmental nurturing information by customers in shopping for products and services. For the present study, pretest interviews with ten retailer shop owner-managers in the TLM supported the conclusion that all these retailers embrace the following perspectives: Because the majority of shops in TLMs are family-owned across multiple generations from premodern to modern times in selling traditional, small-lot size products, such TLMs are high in environmental sustainability; shoppers recognize the TLM as being high in environmental sustainability; and shopper recognition of the TLM as high in environmental sustainability favorably impacts the frequency of their visits and their expenditures in the TLM. The findings of the present study only selectively support the TLM shop owner-managers' perspectives.

Cultural Sustainability Cultural sustainability in retailing refers to preferences for use of traditional products, services, and processes while shopping and using of purchases as well as recognizing and supporting multi-culturalism both across domestic (e.g., tribes and villages) and global origins. Because the majority of shops in the TLM offer a wide variety of products and services that differentiate distinctly different cultural origins mostly within the Korean national culture, the expectation before the present study was that most TLM shoppers would identify the TLM as being high to very high in cultural sustainability and that this perception favorably impacts their visits and expenditures while shopping in the TLM. This perspective received only limited support in the findings of the study.

Wiese et al. (2015) propose that retailers have various means of influencing social, economic, and environmental sustainability. Retailers can improve the sustainability of their products and processes, such as their transports, packaging, and assortment. Retailers can influence the behavior of their supply chain partners and their customers. Wiese et al. (2015) propose that retailers should behave responsibly to maintain their license to operate while at the same time supporting their customers and business partners in behaving more responsibly. Studies assessing retailers' engagement in sustainability include the research by Jones et al. (2011) on the top ten global retailers' reports on their sustainability performances (Wal-Mart, Carrefour, Tesco, Metro, Home Depot, Kroger, Schwarz Unternehmens, Treuhand, Target, Costco, Aldi GMBH). Jones et al. (2011, p. 268) conclude that these top ten global retailers were only following a "weak" rather than a "strong" model of sustainability and their "current policies are little more than genuflections to sustainability." Wilson (2015) provides a useful checklist for an independent management audit of a retailer's engagement in sustainability and the outcomes of such engagements. Wilson's (2015) checklist is an elaboration of Elkington's (1997) "triple bottom line" (economic, environmental, and social) accounting of a firm's performance. Wilson (2015) applies his checklist to assess the triple bottom line performance of Marks and Spencer (M&S)—a global retailer with headquarters in Britain—and he concludes that M&S is engaged heavily in all 15 features in the checklist.

The present study focuses on how customers of a TLM assess the sustainability performance of the TLM. The study examines sustainability for each of four sustainability dimensions, economic, social, environmental, and cultural, as well as testing the proposition taking a macro perspective for sustainability across the four subdimensions. P2 (P2a and P2b) is the proposition that customers having certain demographic identities assess the TLM as providing high performance on sustainability in general (P2a) and/or for specific sustainability dimensions but not all dimensions. The findings below do not support P2a but do support P2b. The findings support the conclusion that sustainability assessment needs to performed for each sustainability dimension and not by taking a global sustainability perspective only.

2.3 P3: Specific Demographic Configurations Exhibit High TLM (Brand) Attractiveness, While Other Specific Demographic Configurations Exhibit Low TLM (Brand) Attractiveness

Figure 5.1 includes P3a visually as an arrow from demographic configurations to brand attractiveness configurations. P3a: A few complex antecedent configurations of demographic characteristics identify shoppers highly attracted to the TLM using a macro measure of TLM. P3b: A few complex antecedent configurations of demographic characteristics identify shoppers highly attracted to the TLM using micro (each of four dimensions) measures of TLM. P3c: A few complex antecedent configurations of demographic characteristics identify shoppers who find the TLM to be unattractive using a macro measure of unattractiveness. P3d: A few complex antecedent configurations of demographic characteristics identify shoppers who find the TLM to be unattractive using micro (each of four) measures of unattractiveness. The third proposition is a rather stark perspective that customers who find a TLM attractive or unattractive at the macro and micro levels are identifiable by demographic profiles alone. The findings below support P3 in general with the exception for P3c. Contrary to P3c, a global positive TLM place attractiveness is not identifiable by a shopper displaying specific alternative demographic profiles.

2.4 P4: TLM Sustainability Configurations and TLM Attractiveness

P4: Market visitors with identifiable configurations of assessments of TLM's sustainability performance indicate high TLM attractiveness. For example, P4 includes the proposition that customers with top quintile scores across all four sustainability dimensions judge the TLM to be high in trust consistently:

$$\text{social} \bullet \text{economic} \bullet \text{culture} \bullet \text{environmental} \;'' \; \text{trust} \qquad (5.6)$$

P4 includes the sub-proposition that market visitors in the top quintile for any three of the four dimensions of TLM sustainability judgments judge the TLM in the top quintile of one or more for place attractiveness. Examples of predictions for trust and image outcomes are:

$$\text{social} \bullet \text{economic} \bullet \text{culture} \;'' \; \text{trust} \qquad (5.7)$$

$$\text{social} \bullet \text{economic} \bullet \text{culture} \;'' \; \text{image} \qquad (5.8)$$

P4 also includes the sub-proposition that market visitors in the bottom quintile consistently in their judgments across all four sustainability dimensions for the

TLM consistently have negative judgments of the attractiveness of the TLM. Equation 5.9 states this sub-proposition for the negation of the four attractiveness dimensions, that is, trust:

$$\sim \text{social} \bullet \sim \text{economic} \bullet \sim \text{culture} \bullet \sim \text{environmental} \leq\sim \text{Trust} \quad (5.9)$$

Simply put, market visitor sustainability judgments about the market should and do matter. Such judgments impact some-to-most visitors' judgments about the market's attractiveness. While P4 may appear to be true at face value, empirical evidence for such consistency at the level of case analyses is lacking, and evidence of such associations at the variable level is scarce (e.g., Wilson 2015).

Figure 5.1 also proposes P4X: The impacts of customer sustainability assessments of a TLM on TLM place attractiveness are relevant for only a few specific demographic configurations of shoppers. P4X expresses the perspective that sustainability assessment explanations of TLM attractiveness is a learned view of older shoppers having completed high education and achieved high income. Concerns with budgetary limitations may limit the concerns with TLM sustainability issue among shoppers with low-income and low education levels. Consequently, the impact of TLM sustainability judgements on place attractiveness is identifiable only for a screened share of shoppers and not for most shoppers. This study tests this theoretical modification of P4.

2.5 P5: Sustainability Assessments and Customer Behavior (Visits and Expenditures)

P5: Customers with high assessments across three to four sustainability conditions for the TLM engage in high visits and high expenditures. Stone's "ethical shopper" informed the crafting of P5. Following Stone (1954), the rationale for P5 is that customers with consistently high sustainability assessments for the TLM are likely to be ethical shoppers. "They perceive shopping in the light of larger set of values rather than of specific values and more immediately relevant norms. Store patronage was appraised [by these ethical shopper] in anticipation of such moral consequences" (Stone 1954, p. 38). Shoppers in the top quintile in assessment of a TLM across each of the four sustainability issues are most likely to perceive that achieving high sustainability matters in large-scale markets. This view supports theoretical models 10–11:

$$\text{social} \bullet \text{economic} \bullet \text{culture} \bullet \text{environmental} \; '' \; \text{visits} \quad (5.10)$$

$$\text{social} \bullet \text{economic} \bullet \text{culture} \bullet \text{environmental} \; '' \; \text{expenditures} \quad (5.11)$$

$$\sim \text{social} \bullet \sim \text{economic} \bullet \sim \text{culture} \bullet \sim \text{environmental} \leq\sim \text{visits} \quad (5.12)$$

$$\sim \text{social} \bullet \sim \text{economic} \bullet \sim \text{culture} \bullet \sim \text{environmental} \leq \sim \text{expenditures} \quad (5.13)$$

Prior theory and research findings do not support the view that negation models (i.e., models 12 and 13) impact visits and expenditures in a TLM. Asymmetric, customer, negation sustainability models identify customers who view a TLM as very low across the four sustainability conditions. Very few such customers are likely to exist. Such customers are unlikely to perceive a connection with their market sustainability perceptions and their own shopping behavior.

P5X in Fig. 5.1 is a sister perspective of P4X. P5X: The impact of shoppers' TLM sustainability assessments on TLM visit intentions and expenditures is limited to a few demographic configurations of shoppers and not across the board of all shoppers. Rationale: The issue of using sustainability assessments to guide one's own behavior is a learned behavior among older, high-income, and highly educated shoppers—not all shoppers. If accurate, sustainability assessment training (SAT) will likely to be a requirement for gaining widespread shoppers' participation in TLM sustainability programs. The study here examines the efficacy of P5X.

2.6 P6: Place Attractiveness and Shopping Behavior (Visits and Expenditures)

Figure 5.1 includes four dimensions of mall place attractiveness: image, satisfaction, loyalty, and trust. P6a: Customers with very high (e.g., top quintiles) scores across all four attractiveness dimensions have very high (membership) scores in mall visits and expenditures. Customers exhibiting such a complex antecedent condition represent a state of "customer delight" (Oliver et al. 1997). This view expands on Oliver et al. (1997) perspective of customer delight as being unexpected high satisfaction to propose that customers assessing a service/shopping experience as consistently very high on several dimensions enter a state of customer delight.

P6b: For very high membership scores in visits and expenditures, customers must have very high scores for at least two dimensions of mall attractiveness. P6b expresses the complexity principle that single antecedent conditions are insufficient for consistently identifying cases having high scores for an outcome condition. P6c: Customers with low scores across all four attractiveness dimensions have very low membership scores in mall visits and expenditures. Such customers with consistently very negative (e.g., bottom quintiles) across service/shopping dimensions represent a state of customer displeasure. P6d: For very low membership scores in visits and expenditures, customers must have very low scores for at least two dimensions of mall attractiveness. Separate asymmetric models useful for indicating high or low scores in an outcome condition rest on the same foundations of complexity theory. Referring to TLM place attractiveness dimensions in Fig. 5.1, here are the core propositions of P6: $P6c_V$, $I \bullet S \bullet L \bullet T \leq V$ and $P6c_E$, $I \bullet S \bullet L \bullet T \leq E$, where V means visits and E expenditures (14), and $P6c_V$, ~I•~S•~L•~T ≤ ~V and $P6c_E$, ~I•~S•~L•~T ≤ E, where ~V few visits and ~E low expenditures (15).

P6x: The impact of TLM place attractiveness on TLM shopping plans and expenditures depends on memberships in specific shopper demographic configurations. Specific shoppers identifiable by their demographic configurations are "crazy in love" with TLM shopping and intend to visit the TLM frequently. Specific shoppers identifiable by their demographic configurations have negative assessments of the TLM that transforms into negative plans to return to the TLM and to spend less if they do return. P6x examines these separate asymmetric modeling perspectives.

2.7 P7: Sustainability Assessments and Place Attractiveness Configurations Explain and Accurately Predict Expenditures and Trip Plans to a TLM

Figure 5.1 includes P7. P7: Shopper configurations by their TLM sustainability assessments and shopper TLM place attractiveness have unique impacts on shopper TLM intentions and expenditures. Shoppers high in both configurations of TLM sustainability assessments and configurations of place attractiveness have high membership scores in planned visits and expenditures. This perspective implies that while such configurations are likely to represent few shoppers, these shoppers are highly supportive in their behavior toward the TLM. This perspective suggests that TLM shoppers' high sustainability assessments extend and complement their high TLM place attraction. The present study examines this extended view of shopper sustainability assessments.

3 Complexity Theory Applications to Place Assessment Research

Complexity theory and data analysis via asymmetric models (i.e., configuration consistency of conditional screens) provides a radically different perspective and analytical steps versus the current dominant logic of linear model construction and data analysis via symmetric models (i.e., reporting of net effects of independent variables on a dependent variable). Complexity theory includes expressing tenets regarding case-based outcomes—tenets that are commonly observable in data. Table 5.2 summarizes five principal complexity theory tenets (Ti).

T1: Insufficiency An antecedent condition (e.g., high trust) may be necessary for the occurrence of a specific outcome condition (e.g., high number of monthly visits), but antecedent conditions are rarely sufficient individually in producing the outcome condition. Configurations consisting of two to a few antecedent conditions are sufficient in indicating the occurrence of the outcome condition. Such configurations represent useful screens that indicate a focal outcome condition.

T2: Equifinality A few configurations are accurate consistently in identifying a focal outcome condition. Using both from ex ante (theory to observation) and abductive (observation to theory) perspectives confirm that more than one solution (i.e., more than one complex antecedent condition) identifies the same outcome (i.e., high membership scores in an outcome condition, such as cases with frequent monthly visits to a TLM).

Related to the equifinality tenet, identifying the relative importance of antecedent conditions is not a central issue in case-based research—though certain simple conditions may appear in several useful configurations more often than other simple conditions, and from this perspective some simple conditions are viewable as more important than others. A consistently accurate, complex, antecedent condition may be viewable as more "important" than another if the former has higher "coverage" than the latter. "Coverage" refers to the share of cases having the focal outcome condition that a complex antecedent condition identifies consistently. For example, if 40 of 200 cases are members of a focal outcome condition and complex antecedent condition labeled, "L," identifies 15 of 40 cases having the focal outcome with 1 inconsistency (inaccuracy), while a second complex antecedent condition labeled "R" identifies 5 of the 40 cases with 1 inconsistency, the L condition may be viewable as more important than the R condition. However, details of both complex antecedent conditions L and R are worth reporting. Accurate consistency is primal in concluding a model is useful. Both models L and R are useful with the findings indicating the odds of accurate identification are 15 to 1 for model L and 5 to 1 for model R. These high odds support the conclusion that equifinality solutions for the focal outcome condition occur. Both models need reporting. One measure of coverage is the number of cases a model predicts that are actually high in an outcome condition. Thus, for model L, coverage is equal to 14/40 (i.e., 0.35), and for model R, coverage is equal to 4/40 (i.e., 0.10).

T3: Contrarian Cases Cases exhibiting independent and dependent variable associations that are contrary to directional main effect relationships occur in nearly all sets of data—even when the correlations among the independent (e.g., T, X, and R) and dependent (e.g., Y and Z) variables are high ($r > 0.60$). Researchers using symmetric tests (e.g., correlation, F-tests, MRA, and SEM) usually ignore contrarian cases and report only statistical significance of directional hypotheses. In embracing complexity theory and asymmetric analysis, the present study includes modeling the configuration where high X associates with high Y, where low X associates with high Y, where high X associates with low Y, and where low X associates with low Y in complex configurations that are consistently accurate in indicating high Y or low Y.

T4: Causal Asymmetry The ingredients (i.e., simple conditions) inside complex statements accurately indicating cases having high scores for an outcome condition differ in content, and some of these conditions differ in direction from the ingredients in other complex statements accurately indicating cases having scores for the

Table 5.2 Core tenets of complexity theory

Tenet	Concept	Description	Boolean Expression
T1	Insufficiency	High X may be necessary, but this condition is insufficient for identifying high Y	X //→ Y
T2 T3	Equifinality Contrarian	A few, not one, distinctly unique complex configurations Of antecedent considered indicate the same outcome Both high X and low X associate with high Y Both high X and low X associate with low Y	(X•R) ≤ Y + (~X•T) ≤ Y (X • R) ≤ Y + (~X • T) ≤ Y (X • W) ≤ ~Y + (~X • F) ≤ ~Y
T4	Causal asymmetry	Complex antecedent conditions for low Y are not the mirror opposite of complex antecedent conditions for high Y	(X • R) ≤ Y ≠ (~X • ~R) ≤ ~Y
T5	Emergence	System effects occurring in creating configurations of simple conditions are greater than the sum of the simple conditions (where SE is self-esteem, GSE generalized self-efficacy, LC locus of control, ES emotional stability, and CSE core self-evaluations)	(SE•GSE•LC•ES > CSE_{total})

Key: Boolean algebra operational meanings: mid-level dot, "•", indicates the logical "and"; sideways tilde, "~", indicates negation; the plus size "+" indicates "or"; the less than or equal sign "≤" indicates scores for the model input statement are all or nearly all lower than scores for the outcome, Y or (Y • Z); the not equal sign "//→" indicates that the input model (simple or complex) does not indicate an asymmetric pattern that screens for Y or ~Y where "Y" refers to cases with high Y scores and "~Y" refers to cases with low Y scores, the negation of a Y score; "X" refers to high X scores and "~X" refers to low X scores. X, R, F, and W refer to simple antecedent conditions; Y and Z refer to simple outcome conditions; "≠" refers to causal asymmetry
Notes: A useful heuristic is to discretize scores when calibrated values of a variable into fuzzy-set scores so that all cases in the lowest quintile have fuzzy-scores ≤0.10 and cases in the highest quintile have fuzzy-set scores ≥0.90. Configural analysis and setting consistency requirements are "fuzzy" in deciding what constitutes low (e.g., ~Y) and high (Y) scores and in deciding on the limit necessary for models of complex antecedent conditions to surpass to indicate high accuracy in predicting Y or ~Y

outcome condition. The causes of success differ in content from the causes of failure. Deep analysis of success versus failure need separate examinations. Thus, Fig. 5.1 shows the present study modeling the causal configurations for V, ~V, E, and ~E.

T5: Emergence System effects occurring in creating configurations of simple conditions are greater than the sum of the simple conditions. For example, using membership scores for four simple conditions (C, R, T, and S) in a complex antecedent condition whereby a 0.85 is pre-set as the minimal score necessary for each condition to predict a high score for an outcome condition (Y), the configuration scores for the first, but not the second, case of the following two cases predict the first case to have a high outcome score. Case 1: $C = 0.85$; $R = 0.91$; $T = 1.00$; $S = 0.86$;

configuration score = 0.85, total score = 3.62. Case 2: $C = 0.94$; $R = 0.98$; $T = 0.99$; $S = 0.78$; configuration score = 0.78, total score = 3.69. Analyses by algorithms are done using configuration scores and not total scores of the simple conditions in configurations. The scores for each simple condition must indicate full membership in the condition for predicting a high membership score in the outcome condition. Non-compensatory decision rules are pre-set in analysis by algorithms. Taking the lowest score across all simple conditions in a complex antecedent statement as the membership score for the entire statement indicates the membership share that each of the conditions in the statement share in common. The lowest score for case 1 in the example equals 0.85; the lowest score for case 2 equals 0.78. Thus, a Boolean and not a matrix algebra operation is used for calculating a case's membership score in a specific complex antecedent configuration.

4 Method

Data were collected on-site at Seomun Market during March 2017 (10 AM to 8 PM) via seven interviewers. The interviewers were paid based on achievement ($5 every hour and $1.00 for each survey completed). Using face-to-face survey to collect data, each survey took about 10 minutes to complete. When respondents completed the survey, the respondents received a "choco pie ($0.5)" as a gift. A total of 162 customers completed the survey (a 70% response rate among persons asked to participate). "Busy, sorry, no time" was the principal reason offered for not agreeing to participate.

The TLM sells a wide range of products. Indeed, 50% report spending less than $200 per month in TLMs, between $200 and $500 (27%), between $500 and $1000 (16%), and between $1001 and $1500 (5%), and five respondents (2%) report spending more than $1500 per month in the TLM. Less than half (43%) report visiting under 3 times per month in TLMs, between 3 times and 6 times (19%), between 7 times and 9 times (10%), and between 10 times and 12 times (5%), and the rest of the respondents visiting over 13 times per month in the TLM.

5 Variables

The variables used in this study were derived through literature reviews and modified to fit the purpose of the study. First, the economic activities among the four sustainable activities are from Arli and Lasmono (2010), Bansal (2005), Carroll (1979), Kim et al. (2012a), Kim et al. (2015a), Maignan (2001), and Maignan and Ferrell (2001). Economic variables were reconstructed based on the measurement items used in the literature. Five items were measured by the five-point Likert scale. Second, the social activities among the four sustainable activities are from Arli and Lasmono (2010), Kim et al. (2015b), Sen and Bhattacharya (2001), and Van Herpen et al. (2003). Social variables were reconstructed based on the measurement items used in the literature. Five items were measured using five-point Likert scale. Third, the environmental activities among the four sustainable activities are from Baughn and McIntosh (2007), Carroll (1979), Kim et al. (2015b), Pivato et al. (2008), and Van Herpen et al. (2003). Environmental variables were reconstructed based on the measurement items used in the literature. Five items were measured by the five-point Likert scale. Last, the cultural activities among the four sustainable activities are from Kim et al. (2015b) and Sen and Bhattacharya (2001). Cultural variables were reconstructed based on the measurement items used in the literature. Five items were measured by the five-point Likert scale.

Brand attractiveness variables are from Ball et al. (2004), Chen (2010), Faullant et al. (2008a, b), Jin et al. (2012a, b), and Nyadzayo and Khajehzadeh (2016). Image variables were reconstructed based on the measurement items used in the literature. Six items were measured by five-point Likert scale items. Satisfaction variables were reconstructed based on the measurement items used in the literature. Four items were measured by five-point Likert scale items. Trust variables were reconstructed based on the measurement items used in the literature. Four items were measured by five-point Likert scale. Loyalty variables were reconstructed based on the measurement items used in the literature. Four items were measured by five-point Likert scale.

6 Validity and Reliability Analysis of Measurement

In this study, Cronbach's alphas were calculated to confirm the reliability of the measurement items using the SPSS 21.0 program and confirmatory factor analysis using AMOS 18.0 program to ensure validity. Reliability and confirmatory factor analysis were conducted to verify reliability and validity of measurement variables. As a result of the confirmatory factor analyses, model fit $\chi^2 = 407.788$, DF = 335, $p = 0.000$, CMIN/DF = 1.873, GFI = 0.861, CFI = 0.929, IFI = 0.930, TLI = 0.920, NFI = 0.861, and RMSEA = 0.057 indicate an appropriate model for study (Hair 2007). Factor loading, significance, average variance extracted, and construct reliability were examined to verify the convergent validity of the measurement model. The factor loading of the measurement variables was statistically significant at 0.6 or more. As a result of calculation of AVE and CR values, the AVE values are 0.5 or more, and CR values are 0.7 or more, and each measurement variable has intensive validity (Hair 2007). Cronbach's α analysis was performed to confirm the reliability of the items. All of them satisfied internal consistency of 0.8 or higher. To verify the discriminant validity of the measurement model, the AVE value of each latent variable was compared with the square of the correlation coefficient between each of the latent variables according to the method proposed by Hair (2007). Table 5.3 includes the correlation coefficient squares and AVE values for each variable.

Table 5.3 Results of confirmatory factor analysis for measurement

Variables		Questionnaires	Factor loading	Cronbach's α	AVE	CR
Sustainable marketing activity	Economic (3)	Efforts are made for efficient management	0.805	0.805	0.591	0.804
		It makes a lot of effort for technological innovation	0.796			
		It improves the economic power of the traditional market with various activities	0.703			
	Social (3)	It supports activities for the community	0.875	0.923	0.633	0.848
		Return some of the profits to society	0.825			
		Part of the amount of profit is devoted to donation	0.701			
	Environmental (3)	Environmentally friendly materials used	0.810	0.877	0.576	0.815
		Consider the environment throughout the design process	0.771			
		Prevent environmental pollution during production and distribution process	0.707			
	Cultural (3)	Respect various races	0.871	0.836	0.610	0.838
		Recognize international multiculturalism	0.817			
		It strives for collaboration with globalization and local culture	0.691			
Brand image (4)		I have a good impression of Seomun ("west gate") market	0.836	0.926	0.555	0.859
		Shoppers have a good image in their minds of Seomun ("west gate") market	0.756			
		It has a better image than other traditional markets	0.687			
		It makes me feel like to visit	0.691			

(continued)

Table 5.3 (continued)

Variables	Questionnaires	Factor loading	Cronbach's α	AVE	CR
Brand satisfaction (3)	I like to visit	0.833	0.843	0.651	0.853
	When I visit, it will be satisfactory	0.823			
	I like to visit and buy things	0.763			
Brand trust (3)	Be honest with the consumer	0.910	0.912	0.679	0.870
	Provide high-quality services and products	0.798			
	I trust this market	0.758			
Brand loyalty (4)	I will choose it than other traditional markets	0.801	0.807	0.593	0.865
	I will invest time and money to visit	0.774			
	I will use it as a traditional market	0.754			
	I would recommend it to others	0.751			

Demographics
What is your age? What is your gender? 1 = male, 0 = female. Are you married? 1 = married, 0 = single What is your job? 1 = retailer, 0 = customer
Academic achievement? ① Up to and graduate high school, ② university course, ③ graduate university, ④ graduate school course, ⑤ finish graduate school

7 Calibrating Variable Values to Case Membership Scores

Configural or set theory methods such as qualitative comparative analysis (QCA) to examine the effects of configurations of conditions (recipes) on outcomes. These configurations represent a "recipe of causally relevant ingredients linked to an outcome" (Ragin 2008, p. 9). Using theory on intersections between sets, Ragin (2008) argues that forces and events (conditions) intersect to cause outcomes and that this is consistent with much of the thinking in qualitative but not the quantitative research in the social sciences. Using Boolean algebra, this intersection is denoted by "logical and" between the condition, and the score for a configuration is equal to the lowest value among the individual conditions in the configuration.

7.1 Data Calibration

Fuzzy-set theory (Zadeh 1965) enables a case's degree of membership in a set to be estimated in fine gradations ranging from full non-membership (0.00) to full membership (1.00). This is achieved by calibrating the value for each case against a standard range of values that represent the accumulated knowledge about a set. The range of values that represent the set is determined by examining information on the entire set (Ragin 2008) to identify the value that represents the threshold full non-membership (set at 0.05); the median as the point of "maximum membership ambiguity" (set at 0.50); and the threshold of full membership (set at 0.95) These three breakpoints represent an externally derived assessment of membership for the set and therefore permit calibration of cases into degrees of membership. Calibrating original scores into fuzzy-set scores can transform ordinal or interval scales into degree of membership in the target set. Ragin (2008) refers to this as "a truth value" that reflects each case's position in an established set rather than just positioning the cases relative to each other as is usual in social science. Calibrating the degree of membership of each case allows comparison of each case against a known external standard that is consistent with existing theory about the set and enables the identification of fine gradations in the degree of membership of a set. Ragin (2008, p.8) emphasizes "All fuzzy sets must be calibrated" and that "it is impossible to conduct meaningful fuzzy set-theoretic analysis without attending to issues of calibration."

The current study applies configural analysis to explore the relationships. The study does this by exploring whether or not the propositions of the study consistently account for nations in truth tables and then determines whether the models are suggested in their full or a more parsimonious form by the fsQCA software (as per Ragin 2008). A truth table shows the possible configurations of causal conditions and the number of cases that are members of those conditions (Ragin 2008). The truth tables for the present study were restricted to report models achieving consistency indexes beyond 0.84 in accurately predicting model high membership outcomes. Also, additional models were estimated to directly test configurations of culture dimensions constructed based upon theory. This procedure is consistent with Fiss (2007) who suggests that specific configurations should be proposed and tested. The study also explores the truth tables and uses fsQCA software to reveal other configurations that are then discussed. The methodology and rationale are outlined in the following sections for each of configural analysis, calibration methods, and results interpretation.

Calibration was performed for all antecedent conditions and the outcome conditions (tipping behaviors) for the data in the study. For each condition, the median value was used for the crossover point of maximum ambiguity equal to 0.50; the threshold for full membership was equal to a score immediately below the highest original scores for 10 percent of the respondents for a variable; the threshold for full non-membership was equal to a score immediately above the lowest original scores for 10 percent of the respondents in the study. The calibration subroutine in the software program (at fsQCA.com) computed the scores for all respondents once the three breakpoints were specified. For all conditions, two to three different sets of breakpoints were set and findings examined; the findings did not indicate substantial differences across the use of different breakpoints.

8 Consistency and Coverage

Consistency is a measure of the degree to which a specific recipe of conditions is consistent with the outcome. A high consistency score indicates that nearly all cases with a high membership score in the recipe of conditions would also have a high score in the outcome condition. High consistency occurs when a case's level of membership in the causal recipe is less than or equal to its level of membership in the outcome (i.e., the causal recipe is a subset of the outcome) (Ragin 2008). The consistency score as calculated in the fsQCA software determines consistency ($X_i \leq Y_i$), where X_i is case i's membership score in set X and Y_i is case i's membership score in the outcome condition, Y, by dividing the sum of the minimum of X_i or Y_i by the sum of X_i. Consistency $(X_i \leq Y_i) = \Sigma \{\min(X_i, Y_i)\} / \Sigma (X_i)$. This approach penalizes consistency indexes for occurrences when X_i exceeds Y_i in proportion to the size of the inconsistency. The present study includes setting a minimum consistency index equal to or greater than 0.85 as indicating high consistency. Achieving a 0.85 consistency index represents a somewhat precise outcome test (SPOT) whereby a substantial majority of cases high in the complex or simple antecedent condition are high in outcome scores—typically the odds are greater than 4 to 1 that cases high in antecedent condition are high in the outcome condition when consistency is equal or greater than 0.85.

The coverage index in fsQCA assesses the degree to which a causal configuration accounts for instances of an outcome. Set theory accommodates the idea that there are alternate causal recipes that lead to the outcome. QCA compares alternate paths by determining the proportion of cases that are consistent with a particular configuration (Ragin 2008). If a configuration explains the outcome for a high versus low proportion of cases, then that configuration is considered more relevant empirically. Coverage should only be considered after high consistency is established. To calculate coverage, the fsQCA software once again calculates the proportion of cases where degree of membership in the configuration is less than the membership in the outcome but this time divides it by the total score for the outcome condition. This coverage calculation estimates the level to which the causal recipe (the subset) covers the outcome (the larger set) (Ragin 2008). Coverage $(X_i \leq Y_i) = \Sigma \{\min(X_i, Y_i)\} / \Sigma (Y_i)$. Ragin (2008) provides a full description of QCA theory and method; the fsQCA software is available at www.fsQCA.com.

9 Findings

The principal focus of the findings section is to report the results for testing the propositions of the study. Following the results for the seven propositions, this finding section also reports the results for the consistency and coverage of the specific models that appear in the prior theory section.

9.1 *Findings for P1: Demographic Configurations Explaining Customer Visits and Expenditures*

No complex demographic configurations were capable of indicating customers who do visit the TLM frequently (Part A in Table 5.4). Table 5.4 includes a few complex demographic configurations that indicate customers who visit the TLM infrequently (Part B in Table 5.4). High visit frequency depends on customer assessments and not simple or complex demographic screens as the results from testing P2–P7 indicate below. Three-of-the-four configuration models in Table 5.4, Part B (predicting customers who visit the TLM infrequently), include single (not married) and young (~age) visitors. For infrequent visitors, no complex configuration models include the simple conditions of marriage or older customers.

Comparing Table 5.4 Parts A and B illustrate the asymmetric and not symmetric nature of associations of complex antecedent conditions with the outcome of frequent versus infrequent TLM shopping visits. While four models are accurate in indicating shoppers who visit infrequently (Table 5.4, Part B), assuming that the mirror opposites of each of these models indicate high shopping frequency would be inaccurate. The findings support the conclusion that all of the theoretical configurations of five demographics in the study are unable to provide consistently accurate predictions of high shopping frequency.

Table 5.5 Parts A and B, respectively, report the findings for demographic models accurate for predicting high versus low monthly expenditures. One model in

Table 5.4 Testing demographic configurations predicting TLM visits and TLM not visits

(A) **Findings for P frequent visits: customers' demographic configurations indicating high local market frequency of visits**			
Model: visits_c = f(gender, marriage, educ_c, income_c, age_c); intermediate solution: frequency cutoff: 1.00; consistency cutoff: 0.85			
Model	**Raw coverage**	**Unique coverage**	**Consistency**
No models provide high consistencies.			
(B) **Findings for P1_C_few visits: customers' demographic configurations indicating low local market frequency of visits**			
Model: not_visit_c = f(age_c, income_c, educ_c, marriage, gender); intermediate frequency cutoff: 1.00; consistency cutoff: 0.89			
Model	**Raw coverage**	**Unique consistency**	
1 ~marriage*~age_cc*~income_c	0.34	0.08	0.93
2 ~marriage*~age_cc*educ_c	0.40	0.04	0.90
3 ~marriage*~gender*educ_c	0.28	0.05	0.85
4 ~gender*~age_cc*educ_c*income_c	0.20	0.05	0.90
Solution coverage: 0.58; solution consistency: 0.87			

Table 5.5 Testing demographic configurations predicting TLM high and TLM low expenditures

(A) **Findings for P1_C_high expenditures: customers' demographic configurations indicating high expenditures on local market**			
Model: exp_c = f(gender, marriage, educ_c, income_c, age_c); intermediate solution: frequency cutoff: 1.00; consistency cutoff: 0.84			
Model	**Raw coverage**	**Unique coverage**	**Consistency**
1 marriage∗~gender∗age_c∗~educ_c∗income_c	0.10	0.10	0.82
Solution coverage: 0.10; solution consistency: 0.82			
(B) **Findings for P1_C_low expenditures: customers' demographic configurations indicating low expenditures on local market**			
Model: not_exp_c = f(age_c, income_c, educ_c, marriage, gender); intermediate frequency cutoff: 1.00; consistency cutoff: 0.82			
Model	**Raw coverage**	**Unique coverage**	**Consistency**
1 ~marriage∗~age_cc∗~income_c	0.34	0.09	0.89
2 ~marriage∗~age_cc∗educ_c	0.41	0.04	0.89
3 ~marriage∗~gender∗educ_c	0.29	0.05	0.84
4 ~gender∗~age_cc∗educ_c∗income_c	0.21	0.04	0.92
Solution coverage: 0.59; solution consistency: 0.85			

Table 5.5 Part A indicates high monthly expenditures: marriage•~gender•age_c•~educ_c•income_c $\leq$ expenditures (i.e., married, female, older, low education, and high income indicate high monthly expenditures in the TLM). This model's consistency is equal to 0.82 and its coverage is equal to 0.10. The XY plot for this model (not shown here; the plot is available from the authors) indicates that the odds are 7 to 1 that shoppers with membership scores 0.5 and above on this model have shopping membership scores above 0.70. Note that high income alone is present in the model but insufficient for accurate prediction of shoppers with high monthly expenditures. An additional telling finding is the limited number of complex demographic models indicating high monthly expenditures. Explaining, describing, and predicting a substantial share of shoppers with high monthly shopping expenditures require more complex models than complex configurations of the five simple demographic conditions.

In Table 5.5 Part B, model 1 includes low-income shoppers, and model 4 includes high-income shoppers. Such a finding that a low and a high score for a simple antecedent condition can identify cases with high scores for an outcome condition supports the contention that the directionality of an XY association depends on the collective influence of the configuration of attributes and not a simple summation of each attribute's influence (cf. Ordanini et al. 2014, pp. 144–145). Such high and low scores for an attribute indicating high scores (as well as low scores) for an outcome occur in many models in the findings to the present study. The relevancy of complexity theory's equifinality tenet is apparent in the findings in Table 5.5 Part B: A few alternative solutions occur for the same outcome—each solution has high consistency though they vary in their coverages.

Not being married (i.e., ~married) and being young (i.e., ~age) and having high formal education (i.e., "educ_c") appear in three of the four models in Table 5.5 Part B. Model 2 in Table 5.5 Part B is the configuration of these three conditions. Young and single shoppers with high education attainment have low monthly expenditures in this TLM. Does this TLM have other customer segments that do generate high monthly expenditures? Table 5.5 Part A identifies one such market which does indicate the mirror opposites to the three conditions: being married, older, and low formal education. The symmetrical nature of these observations occurs infrequently in asymmetric analysis—such symmetrical findings should not be assumed to occur as symmetric tests assume. Performing asymmetric analysis permits answering each question separately and provides independent confirmation of when symmetric relationships occur and when they do not occur. None of the six conditions alone—high income, low income, marriage, single, high education, and low education—are capable of indicating high or low monthly expenditures. Consequently, focusing on reporting net effects the relative importance of net effects, and "critical success factors" is misguided. None of the six conditions just mentioned appears across all four models in Table 5.5 Part B. In this respect, no simple condition is necessary or sufficient for indicating shoppers with low monthly expenditures. Examining complex configurations (models) indicating specific outcomes accurately is necessary.

10 P2 Findings: Demographics and Place Sustainability

Tables 5.6, 5.7, 5.8, and 5.9 present the models that accurately indicate high versus low scores for each of the four sustainability conditions. Models providing high consistencies occur for each of the eight sets of findings. Such a set of findings indicate that demographic configurations offer robust explanations of shopper sustainability assessments of the TLM.

10.1 Customers' Demographic Configurations Indicating High and Low Economic Sustainability Assessment

Table 5.6 Part A includes four models accurate in indicating high economic sustainability assessments of the TLM. Three of the four models include a high-income condition plus additional conditions. Low income is not a condition in any of the models in Table 5.6 Part A. A relevant conclusion here is that demographic configurations offer a reasonable but not high ability in identifying shoppers indicating the TLM has high economic sustainability.

Table 5.6 Demographic configurations indicating high and low economic sustainability assessment

(A) **Findings for P2 cases high in economic: customers' demographic configurations indicating high economic sustainability assessment**			
Model: eco_c = f(marriage, gender, age_c, educ_c, income_c); intermediate solution: frequency cutoff: 1.00; consistency cutoff: 0.81			
Model	**Raw coverage**	**Unique coverage**	**Unique consistency**
1 ~marriage∗gender∗~age_c∗educ_c	0.19	0.17	0.78
2 marriage∗~gender∗educ_c∗income_c	0.16	0.01	0.79
3 ~gender∗age_c∗educ_c∗income_c	0.22	0.08	0.85
4 ~marriage∗gender∗age_c∗~educ_c∗income_c	0.04	0.02	0.91
Solution coverage: 0.58; solution consistency: 0.87			
(B) **Findings for P2 cases low in economic: customers' demographic configurations indicating low economic sustainability assessment**			
Model: n_eco_c = f(marriage, gender, age_c, educ_c, income_c); intermediate frequency cutoff: 1.00; consistency cutoff: 0.87			
Model	**Raw coverage**	**Unique coverage**	**Consistency**
1 marriage∗~gender∗age_c∗~educ_c∗income_c	0.07	0.07	0.87
Solution coverage: 0.07; solution consistency: 0.87			

Table 5.7 Demographic configurations indicating high and social sustainability assessment

(A) **Findings for P2 cases high in social: customers' demographic configurations indicating high social sustainability**			
Model: soci_c = f(marriage, gender, age_c, educ_c, income_c); intermediate solution: frequency cutoff: 1.00; consistency cutoff: 0.82			
Model	**Raw coverage**	**Unique coverage**	**Consistency**
1 ~gender∗educ_c∗income_c	0.35	0.18	0.80
2 ~marriage∗~gender∗~age_c∗~income_c	0.16	0.10	0.76
3 ~marriage∗~age_c∗educ_c∗income_c	0.32	0.12	0.88
4 ~marriage∗gender∗age_c∗~educ_c∗income_c	0.04	0.01	0.84
Solution coverage: 0.61; solution consistency: 0.80			
(B) **Findings for P2 cases low in social: customers' demographic configurations indicating low social sustainability**			
Model: not_soci_c = f(age_c, income_c, educ_c, marriage, gender); intermediate frequency cutoff: 1.00; consistency cutoff: 0.84			
Model	**Raw coverage**	**Unique consistency**	
1 marriage∗~gender∗~age_c∗~educ_c∗~income_c	0.03	0.02	1.00
2 ~marriage∗~gender∗age_c∗educ_c∗~income_c	0.08	0.08	0.84
3 marriage∗~gender∗age_c∗~educ_c∗income_c	0.07	0.07	0.88
Solution coverage: 0.18; solution consistency: 0.88			

Table 5.8 Demographic configurations indicating high and low environment sustainability assessments

(A) **Findings for P2 cases high in environmental: customers' demographic configurations indicating high environmental sustainability assessments**

Model: envi_c = f(marriage, gender, age_c, educ_c, income_c); intermediate solution: frequency cutoff: 1.00; consistency cutoff: 0.84

Model	**Raw coverage**	**Unique consistency**	
1 ~marriage*~age_c*~educ_c*~income_c	0.19	0.06	0.89
2 ~marriage*~gender*~age_c*~income_c	0.17	0.07	0.82
3 ~marriage*gender*age_c*~educ_c*income_c	0.03	0.01	0.84
4 marriage*~gender*~age_c*educ_c*income_c	0.05	0.05	0.90
5 ~marriage*gender*~age_c*educ_c*income_c	0.13	0.07	0.91
6 ~marriage*~gender*age_c*educ_c*income_c	0.07	0.04	0.86

Solution coverage: 0.45; solution consistency: 0.85

(B) **Findings for P2 cases low in environmental: customers' demographic configurations indicating low environmental sustainability assessments**

Model: not_envi_c = f(age_c, income_c, educ_c, marriage, gender); intermediate frequency cutoff: 1.00; consistency cutoff: 0.82

Model	**Raw coverage**	**Unique consistency**	
1 ~marriage*~gender*age_c*educ_c*~income_c	0.09	0.09	0.85
2 ~marriage*gender*age_c*~educ_c*income_c	0.05	0.05	0.82

Solution coverage: 0.14; solution consistency: 0.84

Table 5.9 Demographic configurations indicating high and low culture sustainability assessments

(A) **Findings for P2 cases high in cultural: customers' demographic configurations indicating high culture sustainability assessments**

Model: cul_c = f(marriage, gender, age_c, educ_c, income_c); intermediate solution: frequency cutoff: 1.00; consistency cutoff: 0.93

Model	**Raw coverage**	**Unique coverage**	**Consistency**
1 marriage*~gender*~age_c*~educ_c*~income_c	0.02	0.02	0.94
2 ~marriage*~gender*age_c*educ_c*income_c	0.08	0.08	0.92

Solution coverage: 0.11; solution consistency: 0.93

(B) **Findings for P2 cases low in cultural: customers' demographic configurations indicating low culture sustainability assessments**

Model: not_cul_c = f(age_c, income_c, educ_c, marriage, gender); intermediate frequency cutoff: 1.00; consistency cutoff: 0.84

Model	**Raw coverage**	**Uniquecoverage**	**Consistency**
1 ~marriage*~gender*age_c*educ_c*~income_c	0.07	0.07	0.84
2 ~marriage*gender*age_c*~educ_c*income_c	0.05	0.05	0.86

Solution coverage: 0.12; solution consistency: 0.85

Table 5.6 Part B includes only one model that is accurate in indicating low economic sustainability assessments of the TLM. Notice also that the amount of unique coverage of this one model is equal to 0.07. This amount of coverage is substantially less than the coverage of model in Table 5.6 Part A. These two models (model 1 Part A versus model 1 Part B) each include three of four attributes that are the mirror opposite of one another: single, male, and young in the model 1 Part A and married, female, and old in Part B.

However, these two models in total are not the mirror opposite of each other. Also, older shopper is a condition in two of the models in Table 5.6 Part A. A relevant conclusion is that demographic configurations are useful for identifying some of the cases indicating high as well as low economic sustainability assessments of the TLM—but additional information is necessary for gaining coverage in models of the majority of shoppers for high and low economic sustainability assessments.

10.2 Demographic Configurations Indicating High and Social Sustainability Assessment

Table 5.7 Parts A and B provide the findings for demographic configurations indicating high versus low social sustainability assessments. Both sets of these models provide higher total coverages of shoppers having high as well low social sustainability assessments. Such a finding is viewable as making sense: Social sustainability of the TLM is more relevant to shoppers' immediate well-being than assessments of the TLM's economic viability. Consequently, a greater share of shoppers is likely to be involved in making high and low social versus economic sustainability assessments. The social question has greater gravitas for the shoppers which likely affects definitely favorable as well as unfavorable judgments.

Low income is an ingredient in all three models in Table 5.7 Part B. Low income also appears in one of the models in Table 5.7 Part A. High income appears in three of the four models in Part A. These findings support the conclusion that low income is local generalizable necessity condition in predicting shoppers with low social sustainability assessments, but low income is an insufficient condition. In a few cases, low income with additional attributes may indicate high social sustainability.

The models in Table 5.7 Part A for high social sustainability assessments provide high total overall coverage (0.61). Coverage is low for the models in Table 5.7 Part B. (0.18) for low social sustainability assessments. These findings indicate high relevancy of demographic configurations in indicating high but not low social sustainability assessments for the TLM.

10.3 Demographic Configurations Indicating High and Low Environment Sustainability Assessments

Table 5.8 indicates that the demographic configurations indicating high (as well as low) TLM environment sustainability assessments are complex. All models in Table 5.8 include at least four attributes. Being married and not married, old and young, male and female, high and low education, and high and low income appear among the models for both high and low TLM environment sustainability assessment outcomes—all models in Table 5.8 have consistently high accuracies. These findings support the conclusion that reporting main and two-way interaction effects of demographic variables for customer judgments of TLM environment sustainability would not provide substantive information on shoppers judging the TLM as being or not being environmentally sustainable.

In Table 5.8, the solution coverage (0.45) for the models in Part A is substantially higher than the solution coverage (0.14) for Part B. This pattern repeats for two of the other three sustainability dimensions. Possible this pattern provides supports for a general observation that greater shares of persons expressing a pro- versus anti-sustainability assessment are identifiable by complex demographic configurations.

10.4 Demographic Configurations Indicating High and Low Culture Sustainability Assessments

Table 5.9 reports the findings for models indicating high accuracy of shoppers for TLM high and shoppers with TLM low culture sustainability assessments. In Table 5.9 Part A, only one attribute is both models: Female appears in both models indicating high culture sustainability assessments. The two models in Part A include five attributes. Demographic configurations are useful for accurate indication of highly positive culture sustainability assessments, but such configurations do not provide high total solution coverage. Solution coverage is only 0.11 in Part A.

The models in Table 5.9 Part B indicate that some shoppers assessing the TLM to be low in culture sustainability are identifiable by complex demographic configurations. The two models in Part B are quite distinct from one another; both models only include one attribute in common: older shopper. Both models include five demographic attributes. These findings support the conclusion that complex demographic configurations are helpful in identifying a few of the shoppers having negative judgments for TLM culture sustainability; however, researchers need to go beyond modeling using only demographic attributes for achieving high coverage of such shoppers. The solution coverage for negative culture sustainability models is only 0.12.

10.5 Demographic Configurations Indicating High and Low Sustainability Across All Four Dimensions

Using complex demographic configurations, the analysis includes explaining and predicting shoppers with high and those low in overall sustainability, that is, shoppers with top quintile (or bottom quintile scores) for all four dimensions of sustainability. Table 5.10 reports the findings. Models for complex demographic configurations were unable to predict shoppers having consistently high scores across all four sustainability dimensions. This finding supports the perspective that shoppers assess each of the four sustainability dimensions uniquely for high sustainability.

Table 5.10 Part B findings do include models of complex demographic configurations that are able to accurately predict highly negative scores across all four sustainability dimensions. Note in Table 5.10 that the models are indeed complex; all six models include four or five attributes, and high scores in either the positive or negative dimension of each demographic condition appear in the models. The overall coverage solution for these findings is high (0.55). These findings illustrate the examination of complex outcome conditions by complex antecedent conditions and present an innovative report on how complex demographic configurations can accurately predict negative overall sustainability assessments. These negative sustainability assessments are complex: Both high and low membership scores occur for each of the five demographics among the six models in Table 5.10 Part B.

Table 5.10 Demographic configurations indicating high and low membership scores for complex demographic outcomes: Across all four sustainability conditions

(A) **Predicting high shopper membership across all four sustainability conditions**			
Model: Overall sustainability = f(marriage, gender, age_c, educ_c, income_c)			
None of the demographic models are accurate in predicting high scores across all four sustainability conditions.			
(B) **Predicting high membership scores in the negation for all four sustainability conditions**			
Model: Negation of overall sustainability = f(marriage, gender, age_c, educ_c, income_c)			
Frequency cutoff: 1.00, consistency cutoff: 0.87			
Model	**Raw coverage**	**Unique coverage**	**Consistency**
1 ~gender∗~age_c∗~educ_c∗~income_c	0.09	0.07	0.90
2 ~marriage∗gender∗~age_c∗~income_c	0.18	0.15	0.83
3 marriage∗~gender∗age_c∗~educ_c	0.11	0.08	0.84
4 ~gender∗age_c∗educ_c∗~income_c	0.18	0.12	0.91
5 ~marriage∗gender∗age_c∗~educ_c∗income_c	0.03	0.00	0.88
6 marriage∗~gender∗~age_c∗educ_c∗income_c	0.05	0.02	0.87
Solution coverage: 0.55; solution consistency: 0.86.			

11 P3 Findings: Complex Demographic Configurations and Place Attractiveness Assessments

Tables 5.11, 5.12, 5.13, 5.14, and 5.15 present the findings for tests of demographic configurations indicating high as well as low membership scores for the four TLM attractiveness attributes. The four attractiveness attributes include shoppers' TLM image, visit satisfaction, trust, and loyalty.

11.1 Demographic Configurations Indicating Cases with High Versus Low TLM Image

Table 5.11 Part A includes five models for cases having highly positively images of the TLM. These models are quite complex: Each model includes four-to-five demographic conditions. Being married or not and being male or female appear in all four combinations of these two conditions in Table 5.11 Part A. Information on marital state and gender is insufficient for indicating consistently high scores on image. Model 2 (marriage•~gender•age•income ≤ visits) illustrates a configuration of four attributes that indicate shoppers having a consistently favorable image of the TLM: older, married, females with high incomes.

Table 5.11 Findings for P3: shopper demographic configurations predicting image assessments

(A) **Findings for P3_C_cases high in image: shoppers' demographic configurations indicating high image assessment**			
Model: image_c = f(marriage, gender, age_c, educ_c, income_c); intermediate solution: frequency cutoff: 1.00; consistency cutoff: 0.82			
Model	**Raw coverage**	**Unique coverage**	**Consistency**
1 ~marriage∗gender∗~age_c∗~income_c	0.19	0.17	0.77
2 marriage∗~gender∗age_c∗income_c	0.18	0.02	0.86
3 marriage∗~gender∗educ_c∗income_c	0.17	0.01	0.83
4 marriage∗~gender∗~age_c∗~educ_c∗~income_c	0.02	0.02	0.89
5 ~marriage∗gender∗age_c∗~educ_c∗income_c	0.04	0.01	0.85
Solution coverage: 0.41; solution consistency: 0.80			
(B) **Findings for P3_C_cases low in image: shoppers' demographic configurations indicating low image assessment**			
Model: negative_image_c = f(marriage, gender, age_c, educ_c, income_c); intermediate solution: frequency cutoff: 1.00; consistency cutoff: 0.81			
Model	**Raw coverage**	**Unique coverage**	**Consistency**
1 ~marriage∗~gender∗age_c∗educ_c	0.11	0.11	0.80
2 marriage∗~gender∗~age_c∗educ_c∗income_c	0.06	0.06	0.81
Solution coverage: 0.18; solution consistency: 0.80			

Table 5.12 Findings for P3: shopper demographic configurations predicting satisfaction assessments

(A) **Findings for P3_C_cases high in satisfaction: shoppers' demographic configurations indicating high satisfaction**			
Model: satis_c = f(marriage, gender, age_c, educ_c, income_c); intermediate solution: frequency cutoff: 1.00; consistency cutoff: 0.86			
Model	**Raw coverage**	**Unique coverage**	**Consistency**
1 marriage∗~gender∗~age_c∗~educ_c∗~income_c	0.02	0.02	0.86
2 ~marriage∗gender∗age_c∗~educ_c∗income_c	0.04	0.04	0.82
Solution coverage: 0.07; solution consistency: 0.86			
(B) **Findings for P3_C_cases low in satisfaction: shoppers' demographic configurations indicating low satisfaction**			
Model: negative_sati_c = f(marriage, gender, age_c, educ_c, income_c); intermediate solution: frequency cutoff: 1.00; consistency cutoff: 0.81			
Model	**Raw coverage**	**Unique coverage**	**Consistency**
1 ~marriage∗~age_c∗~educ_c∗~income_c	0.23	0.23	0.83
Solution coverage: 0.23; solution consistency: 0.83			

Table 5.13 Findings for P3: shopper demographic configurations predicting trust assessments

(A) **Findings for P3: cases high in trust: shoppers' demographic configurations indicating high TLM trust assessment**			
Model: trust_c = f(marriage, gender, age_cc, educ_c, income_c); intermediate solution: frequency cutoff: 1.00; consistency cutoff: 0.85			
Model	**Raw coverage**	**Unique coverage**	**Consistency**
1 ~marriage∗gender∗~age_c∗educ_c	0.21	0.19	0.82
2 marriage∗~gender∗~age_c∗~educ_c∗~income_c	0.02	0.02	1.00
3 ~marriage∗gender∗age_c∗~educ_c∗income_c	0.04	0.02	0.86
Solution coverage: 0.25; solution consistency: 0.83			
(B) **Findings for P3: cases low in loyalty: shoppers' demographic configurations indicating low TLM trust assessment**			
Model: negative_loy_c = f(marriage, gender, age_cc, educ_c, income_c); intermediate solution: frequency cutoff: 1.00; consistency cutoff: 0.86			
Model	**Raw coverage**	**Unique coverage**	**Consistency**
1 ~marriage∗~gender∗~age_c∗~educ_c∗~income_c	0.10	0.08	0.86
2 ~marriage∗~gender∗age_c∗educ_c∗~income_c	0.08	0.07	0.98
3 marriage∗~gender∗~age_c∗educ_c∗income_c	0.07	0.07	0.88
Solution coverage: 0.23; solution consistency: 0.90			

Table 5.14 Findings for P3: shopper demographic configurations predicting loyalty assessments

(A) **Findings for P3_C_cases high in loyalty: shoppers' demographic configurations indicating high loyalty**			
Model: loy_c = f(marriage, gender, age_cc, educ_c, income_c); intermediate solution: frequency cutoff: 1.00; consistency cutoff: 0.86			
Model	**Raw coverage**	**Unique coverage**	**Consistency**
1 marriage∗~gender∗~age_c∗educ_c∗income_c	0.07	0.07	0.86
Solution coverage: 0.07; solution consistency: 0.86			
(B) **Findings for P3_C_cases low in loyalty: shoppers' demographic configurations indicating low loyalty**			
Model: negative_loy_c = f(marriage, gender, age_cc, educ_c, income_c); intermediate solution: frequency cutoff: 1.00; consistency cutoff: 0.80			
Model	**Raw coverage**	**Unique coverage**	**Consistency**
1 ~marriage∗~age_c∗~educ_c∗~income_c	0.22	0.15	0.87
2 ~marriage∗~gender∗educ_c∗~income_c	0.19	0.03	0.88
3 ~marriage∗~gender∗~age_cc∗educ_c	0.28	0.12	0.80
Solution coverage: 0.45; solution consistency: 0.81			

Table 5.15 Shopper demographic configurations predicting overall place attractiveness across the four attractiveness dimensions

(A) **Findings for P3: shoppers demographic configurations indicating high market attractiveness across four dimensions**			
Model: mkt_att_c = f(marriage, gender, age_c, educ_c, income_c), intermediate solution, frequency cutoff: 1.00; consistency cutoff: 0.84			
No models found indicating high accuracy for overall market attractiveness.			
(B) **Findings for P3: shoppers' demographic configurations indicating high market unattractiveness**			
Model: n_mkt_att_c = f(marriage, gender, age_c, educ_c, income_c), intermediate solution, frequency cutoff: 1.00; consistency cutoff: 0.84			
Model	**Raw coverage**	**Unique coverage**	**Consistency**
1 ~marriage∗~age_c∗~income_c	0.33	0.07	0.90
2 ~marriage∗~age_c∗educ_c	0.39	0.03	0.87
3 ~marriage∗~gender∗educ_c	0.29	0.04	0.86
4 ~gender∗age_c∗educ_c∗~income_c	0.17	0.08	0.94
5 ~gender∗~age_c∗educ_c∗income_c	0.20	0.02	0.91
6 ~marriage∗gender∗age_c∗~educ_c∗income_c	0.03	0.00	0.87
Solution coverage: 0.66; solution consistency: 0.85			

In the models for cases indicating consistently low images for the TLM, note that not married females are ingredients in model 2 in Table 5.11, Part B—the same two ingredients in model 2 Table 5.11 Part A. Several (i.e., 4-to-5) conditions in complex statements "need to fall into place" for a model to achieve high accuracy in predicting a specific outcome consistently.

The two models in Table 5.11 Part B do not include low income as an ingredient. Model 2 does include high income, and income level is irrelevant in model 1. Income level alone is not a necessary or sufficient condition indicating high or low image in models of shoppers. This statement applies in general to the other four demographic attributes as well.

11.2 Demographic Configurations Indicating Cases with High Versus Low TLM Satisfaction

In two sets of models, Table 5.12 presents the findings for demographic configurations indicate high versus low TLM satisfaction. The findings are sparse: two models useful for shoppers with high satisfaction and one model for shoppers with low satisfaction.

The two models for high satisfaction have only one attribute in common: low education. Both of these are complex; each has five attributes. Coverage is only 0.07 for the two models. These findings support the conclusion that for identifying most shoppers with high satisfaction with visiting the TLM, additional information beyond their demographic configurations is necessary. This additional information likely includes assessments of the TLM facilities/environment and the shoppers' assessments of service quality of businesses in the mall.

In Table 5.12 B, shopper "unmarried, young, low income, with low education" is the one screen providing consistently high accuracy in identifying unsatisfied with visiting the TLM. This finding may confirm expectations held by retailing researchers. Even with low incomes, most of these shoppers have cell phones, and communicating with friends occurs in physical contexts more conducive to texting and sending photographs than in the TLM context.

11.3 Demographic Configurations Indicating Cases with High Versus Low TLM Trust Assessments

Table 5.13 A and B include the models for identifying shoppers with high versus low TLM trust assessments. Both sets of models are quite complex—they include four or five attributes. Note that same attribute does not appear across all three models in Table 5.13 A. For example, models 1 and 3 include unmarried shoppers, and model 2 includes married shoppers. Such findings support the general conclusion that researchers need to craft models beyond net effects and main directional null hypothesis significance tests.

Model 1 in Table 5.13 B is an extended form of model 1 in Table 5.12 B. Both of these models include unmarried, young, low education, and low income. The model in Table 5.12 B does not gender—meaning that the model is applicable for both males and females; however, by including females in the model, the model in Table 5.13 B is more restrictive. Note in Table 13 B the same attribute is included across all three models (e.g., female appears among the attributes in the models). Female shoppers are an attribute that is locally generalizable for low trust assessments in configuration models for this outcome. However, most females do not assign low trust assessments—being female alone is insufficient for identifying low or high trust assessments.

11.4 Demographic Configurations Indicating Cases with High Versus Low TLM Loyalty Assessments

Table 5.14 includes the models indicating shoppers with high versus TLM loyalty assessments. In Table 5.14 A, only one demographic configuration model provides identifies shoppers having high TLM loyalty: married, female, young, high education, and high-income shoppers. In particular, these shoppers are customers that the TLM are likely to be most successful in nurturing long-term relationships.

Note that unmarried appears in all three models of Table 5.14 B, and the highly non-loyal shoppers and female shoppers appear in two of the three models. Note that the model that includes unmarried, young, low education, and low income (sometimes restricted to female) appears in all four outcome conditions for the negations of attractiveness. This model is locally generalizable across the individual outcome dimensions for attractiveness assessments.

12 Demographic Configurations Indicating Cases with High Versus Low TLM for General Attractiveness Assessments

In Table 5.15 A, the findings indicate that no demographic configuration is capable of identifying shoppers with high assessments across all four attractiveness attributes. In Table 5.15 B, the findings include six demographic configuration models that do identify shoppers who have a consistently negative assessment for all four attractiveness dimensions. Being unmarried appears in four of these six models, and being unmarried is a condition in models that qualify shoppers in all six models.

These findings for overall attractiveness indicate that overall high attractiveness is a tougher objective to accomplish in comparison to overall negative attractiveness via complex antecedent configurations. A macro perspective of shoppers with negative overall attraction exists, but a macro perspective of shoppers with positive overall attraction does not. Complex demographic conditions do indicate individual attractiveness conditions and are likely to indicate configurations of two simple attractiveness conditions.

13 P4 Findings: Complex Sustainability Conditions Indicating TLM Attractiveness

Table 5.16 provides the findings from testing P4. From Table 5.16A, no complex simple or complex configurations of sustainability conditions indicate high membership scores in overall TLM attractiveness. From Table 5.16, four models indicate high consistencies in predicting the negation of TLM attractiveness. Note that all four models are not very complex—they each contain one or two conditions. The coverages of these models are quite high.

These findings support the perspective that complex sustainability antecedent models are unable to indicate shoppers having favorable sustainability scores—high sustainability is insufficient for indicating high overall attractiveness. However, shoppers in different complex negation sustainability conditions do hold a negative overall attractiveness toward the TLM. Shoppers having negative sustainability perspectives have negative attractiveness perspectives. Stated colloquially, the perspective receives support that shoppers holding a high TLM sustainability beliefs alone is not sufficient for TLM attractiveness, but shoppers with negative TLM sustainability is sufficient for nurturing negative attractiveness toward the TLM.

Table 5.16 Findings for P4 complex sustainability conditions indicating TLM attractiveness

(A) **Findings for P4_C_high Mkt ATT: shoppers' sustainability dimensions configurations indicating TLM high attractiveness**			
Model: mkt_att_c = f(cult_c, env_c, soci_c, eco_c); intermediate solution: frequency cutoff: 1.00; consistency cutoff: 0.80			
Model	**Raw coverage**	**Unique coverage**	**Consistency**
No models with high consistencies.			
(B) **Findings for P4_C_low mkt ATT: shoppers' sustainability dimensions configurations indicating negation of TLM attractiveness**			
Model: not_mkt_att_c = f(cult_c, env_c, soci_c, eco_c); intermediate solution: frequency cutoff: 1.00; consistency cutoff: 0.91			
Model	**Raw coverage**	**Unique coverage**	**Consistency**
1 ~soci_c	0.52	0.05	0.91
2 ~cult_c	0.59	0.09	0.90
3 env_c*~eco_c	0.38	0.04	0.92
4 ~env_c*eco_c	0.30	0.02	0.95
Solution coverage: 0.75; solution consistency: 0.87			

14 P4X Findings: Sustainability and Demographic Configurations Indicating TLM Market Attractiveness

Table 5.17 includes the findings for testing P4X—adding demographic conditions with sustainability conditions clarifies and deepens explanations for high TLM attractiveness and the negation TLM attractiveness. The findings support P4X. Note that the models in Table 5.17A include three-to-four sustainability dimensions as well as demographic conditions. Now high sustainability conditions are found to matter in indicating shoppers with high attractiveness toward the TLM. All three of the models in Table 5.17A include older shoppers AND low education plus additional demographics. Adding demographics to the description of shoppers for identifying sustainability indicating high TLM attractiveness indicates that sustainability does matter for generating high attractiveness—but such impact occurs for fine (i.e., total coverage is 0.12) segments of shoppers.

Table 5.17B broadens and deepens understanding about the shoppers where connections occur for sustainability and TLM negation attractiveness occurs. Table 5.17B includes 16 models. Most of these models include one or more negations of sustainability conditions. However, a few of the 16 models include only positive conditions for sustainability dimensions.

This finding indicates that some shoppers perceiving the TLM is high in sustainability results in a negative attraction toward the TLM. Not all shoppers see being high sustainability dimensions as favorable for TLM attractiveness—in fact, high sustainability is a turnoff for some shoppers. Models 1 and 7 in Table 5.17B include unmarried AND young AND female shoppers perceiving the TLM as high on sustainability dimensions resulting in negative attractiveness toward the TLM.

15 Findings for P5 and P5X: Customers' Sustainability Configurations Indicating High Versus Low Local Market Frequency of Visits—Without and with Demographic Configurations

The findings do not support P5. Customers' sustainability configurations alone do not indicate high shopping visit frequency. However, the findings do provide some support for P5X: The addition of demographic configurations with sustainability dimensions does enable the identification of shoppers reporting high visit frequencies. Table 5.18A presents these findings. The findings include two models. Both models include high scores in all four sustainability dimensions; model 1 includes five demographic and model 2 includes four demographic configurations. The demographic models are unique in these models except that both models include older, low educated shoppers.

Table 5.17 P4X: Sustainability and demographic configurations indicating TLM market attractiveness

Findings for P4X_C_high Mkt ATT: shoppers' sustainability dimensions and demographic configurations indicating high market attractiveness			
Model: mkt_att_c = f(marriage, gender, age_cc, educ_c, income_c, cult_c, env_c, soci_c, eco_c); intermediate solution: frequency cutoff: 1.00; consistency cutoff: 0.87			
Model	**Raw coverage**	**Unique coverage**	**Consistency**
1 marriage*~gender*age_cc*~educ_c*~income_c*env_c*soci_c*eco_c	0.06	0.01	0.91
2 marriage*~gender*age_cc*~educ_c*cult_c*env_c*soci_c*eco_c	0.06	0.01	0.95
3 ~marriage*gender*age_cc*~educ_c*income_c*cult_c*env_c*soci_c*eco_c	0.05	0.05	0.87
Solution coverage: 0.12; solution consistency: 0.90			
Findings for P4X_C_low Mkt ATT: shoppers' sustainability dimensions and demographic configurations indicating high market unattractiveness			
Model: not_mkt_att_c = f(marriage, gender, age_c, educ_c, income_c, cult_c, env_c, soci_c, eco_c); intermediate solution: frequency cutoff: 1.00; consistency cutoff: 0.90			
Model	**Raw coverage**	**Unique coverage**	**Consistency**
1 cult_c*env_c*soci_c*~marriage*~gender*~age_cc*educ_c	0.15	0.04	0.92
2 ~cult_c*~env_c*~soci_c*~eco_c*marriage*~gender*age_cc*~income_c	0.09	0.03	0.95
3 ~cult_c*env_c*~soci_c*~eco_c*~marriage*~gender*~age_cc*educ_c	0.07	0.01	0.98
4 ~cult_c*~env_c*~soci_c*~eco_c*~gender*age_cc*educ_c*~income_c	0.09	0.01	0.95
5 ~cult_c*~env_c*~soci_c*marriage*~gender*age_cc*educ_c*~income_c	0.07	0.01	0.94
6 env_c*soci_c*eco_c*~marriage*gender*~age_cc*educ_c*~income_c	0.09	0.05	0.97
7 cult_c*env_c*eco_c*~marriage*~gender*~age_cc*educ_c*income_c	0.10	0.01	0.93
8 cult_c*env_c*soci_c*eco_c*~marriage*~gender*educ_c*income_c	0.12	0.01	0.90
9 ~cult_c*~env_c*~soci_c*~eco_c*~marriage*gender*~age_cc*educ_c*~income_c	0.06	0.02	0.96
10 ~cult_c*~env_c*soci_c*eco_c*~marriage*~gender*~age_cc*educ_c*income_c	0.05	0.00	1.00

11 ~cult_c*env_c*soci_c*~eco_c*marriage*~gender*~age_cc*educ_c*income_c	0.04	0.01	1.00
12 ~cult_c*env_c*~soci_c*eco_c*marriage*~gender*~age_cc*educ_c*income_c	0.04	0.00	1.00
13 ~cult_c*~env_c*~soci_c*~eco_c*marriage*gender*age_cc*educ_c*income_c	0.02	0.02	1.00
14 cult_c*env_c*~soci_c*eco_c*marriage*~gender*age_cc*educ_c*~income_c	0.05	0.01	0.96
15 cult_c*~env_c*soci_c*eco_c*marriage*~gender*age_cc*educ_c*income_c	0.04	0.01	0.90
Solution coverage: 0.47; solution consistency: 0.94			

Table 5.18 Frequent visit findings

(A) **Findings for P5X_C_frequent visits: customers' sustainability dimensions and demographic configurations indicating high local market frequency of visits**			
Model: visits_c = f(age_c, income_c, marriage, gender, educ_c, cult_c, env_c, soci_c, eco_c); intermediate solution: frequency cutoff: 1.00; consistency cutoff: 0.81			
Model	**Raw coverage**	**Unique coverage**	**Consistency**
1 cult_c*env_c*soci_c*eco_c*marriage*~gender*age_cc*~educ_c	0.05	0.05	0.81
2 cult_c*env_c*soci_c*eco_c*~marriage*gender*age_cc*~educ_c*income_c	0.05	0.05	0.81
Solution coverage: 0.10; solution consistency: 0.81			
(B) **Findings for P5X_C_few visits: customers' sustainability dimensions and demographic configurations indicating low local market frequency of visits**			
Model: not_visit_c = f(age_c, income_c, marriage, gender, educ_c, cult_c, env_c, soci_c, eco_c) intermediate solution: frequency cutoff: 1.00; consistency cutoff: 0.99			
Model	**Raw coverage**	**Unique coverage**	**Consistency**
1 ~cult_c*env_c*~soci_c*~eco_c*~marriage*~gender*~age_cc*educ_c	0.07	0.02	1.00
2 cult_c*env_c*soci_c*~marriage*~gender*~age_cc*educ_c*~income_c	0.09	0.04	1.00
3 env_c*soci_c*eco_c*~marriage*gender*~age_cc*educ_c*~income_c	0.09	0.05	1.00
4 ~cult_c*~env_c*~soci_c*~eco_c*~marriage*gender*~age_cc*educ_c*~income_c	0.06	0.02	1.00
5 ~cult_c*~env_c*soci_c*eco_c*~marriage*~gender*~age_cc*educ_c*income_c	0.05	0.01	0.99
6 cult_c*env_c*~soci_c*eco_c*~marriage*~gender*~age_cc*educ_c*income_c	0.05	0.01	0.99
7 ~cult_c*env_c*soci_c*~eco_c*marriage*~gender*~age_cc*educ_c*income_c	0.03	0.03	0.99
8 cult_c*env_c*soci_c*eco_c*~marriage*~gender*age_cc*educ_c*income_c	0.05	0.02	1.00
Solution coverage: 0.31; solution consistency: 1.00			

Table 5.18B includes the findings for sustainability plus demographic configurations indicating the negation of visiting the TLM. Seven of the eight models in Table 5.18B include young shoppers. Six of the eight models include females. Two of these models include all high scores in the four sustainability dimensions—six models include one or more negations of the sustainability dimensions. The models in Table 5.18B result in substantially higher coverage (0.31) than the models in Table 5.18A (coverage 0.10). These findings support the following conclusions: Complex sustainability configurations that include demographic attributes do indicate not visiting the TLM. For some shoppers, their high membership scores in sustainability contribute to this negation. High TLM sustainability sometimes contributes to negative visitation.

16 Findings for P5 and P5X for Expenditures: Customers' Sustainability Configurations Indicating High Versus Low Expenditures—Without and with Demographic Configurations

Tables 5.19 and 5.20 present the findings for sustainability configurations and high versus low expenditures in the TLM—first without the addition of demographic attributes (Table 5.19) and with demographic attributes (Table 5.20). The findings have the same pattern as the pattern for visits in that no sustainability configurations alone indicate high TLM expenditures (Table 5.19A). However, specific sustainability configurations alone do indicate the negation of TLM expenditures (Table 5.19B). These models include just one or two sustainability conditions. The negation of culture alone is sufficient for indicating negation of TLM expenditures. The five sustainability models in Table 5.19B achieve a total coverage equal to 0.65. All five models have one or more negative sustainability dimensions. Conclusion: TLM visitors viewing the TLM as low in certain configurations of sustainability indicates low expenditures.

Table 5.20A includes two sustainability plus demographic configurations indicating high expenditures in the TLM. All four sustainability dimensions appear in both models as well as high income, old, and low education plus additional demographic dimensions. These models present complex stories indicating that the TLM offers high sustainability for some shoppers having high expenditures in the TLM.

The first three models in Table 5.20B include positive sustainability dimensions that, along with demographic attributes, indicate the negation of TLM expenditures. Unmarried, female, and young are attributes in all three of these models. These findings offer some credence that unmarried, female, and young shoppers believe that the TLM achieves high sustainability and their belief associates with low to no expenditures at the TLM.

Table 5.19 Findings for P5 expenditures: customers' sustainability dimension configurations indicating high expenditures on local market

(A) **Findings for P5 Negation of expenditures: customers' sustainability dimensions configurations indicating low expenditures on local market**			
Model: exp_c = f(cult_c, env_c, soci_c, eco_c); intermediate solution: frequency cutoff: 1.00; consistency cutoff: 0.87			
Model	**Raw coverage**	**Unique coverage**	**Consistency**
No models with high consistencies.			
(B) **Findings for P5 negation of expenditures: customers' sustainability dimensions configurations indicating low expenditures on local market**			
Model: not_exp_c = f(cult_c, env_c, soci_c, eco_c); intermediate solution: frequency cutoff: 1.00; consistency cutoff: 0.81			
Model	**Raw coverage**	**Unique coverage**	**Consistency**
1 ~cult_c	0.53	0.12	0.80
2 ~env_c*~soci_c	0.34	0.01	0.81
3 ~soci_c*eco_c	0.27	0.02	0.80
4 ~env_c*eco_c	0.27	0.02	0.84
5 env_c*soci_c*~eco_c	0.28	0.03	0.84
Solution coverage: 0.65; solution consistency: 0.79			

17 P6 Findings: Customers' Brand Attractiveness Configurations Without and with Demographic Conditions Indicating High Versus Low Local Market Frequency of Visits

Tables 5.21 and 5.22 provide the findings for P6. The same pattern holds for attractiveness as the findings for sustainability describe. As Table 5.21A indicates, complex attractiveness configurations alone are insufficient for indicating high TLM visit frequency—no models are useful for identifying shoppers with high visit frequency. However, in Table 5.21B reports, seven models of brand attractiveness are useful for identifying shoppers having low TLM visits. These seven models all include one or two negative brand attractiveness dimensions. Being not loyal alone consistently indicates shoppers reporting visiting the TLM infrequently (model 1). While this model makes sense intuitively, its occurrence is suggestive of high nomological validity of the additional models in Table 5.21B. These six additional models are additional solutions for indicating high memberships in not visiting the TLM. All six of these models include some combinations of either trust or not trust and satisfaction or not satisfaction and image and not image. Such nuanced models represent a call for vigilance—high satisfaction or high trust or highly favorable image in the TLM may be present in some shoppers' minds who do *not* visit the TLM.

Table 5.22A includes the models of brand attractiveness with demographic dimensions that indicate high visit frequency. Three useful models indicating high

Table 5.20 High Expenditure findings

(A) **Findings for P5X_C_high expenditures: customers' sustainability dimensions and demographic configurations indicating high expenditures on local market**			
Model: exp_c = f(age_c, income_c, marriage, gender, educ_c, cult_c, env_c, soci_c, eco_c); intermediate solution: frequency cutoff: 1.00; consistency cutoff: 0.82			
Model	**Raw coverage**	**Unique coverage**	**Consistency**
1 cult_c*env_c*soci_c*eco_c*~educ_c*income_c*marriage*~gender*age_c	0.03	0.03	0.98
2 cult_c*env_c*soci_c*eco_c*~educ_c*income_c*~marriage*gender*age_c	0.05	0.05	0.82
Solution coverage: 0.08; solution consistency: 0.88			
(B) **Findings for P5X_C_low expenditures: customers' sustainability dimensions and demographic configurations indicating low expenditures on local market**			
Model: not_exp_c = f(age_c, income_c, marriage, gender, educ_c, cult_c, env_c, soci_c, eco_c) intermediate solution: frequency cutoff: 1.00; consistency cutoff: 0.91			
Model	**Raw coverage**	**Unique coverage**	**Consistency**
1 cult_c*env_c*soci_c*eco_c*~marriage*~gender*~age_cc*educ_c	0.13	0.03	0.93
2 cult_c*env_c*soci_c*~marriage*~gender*~age_cc*educ_c*income_c	0.11	0.01	0.96
3 cult_c*env_c*eco_c*~marriage*~gender*~age_cc*educ_c*income_c	0.11	0.01	0.96
4 ~cult_c*~env_c*~soci_c*~eco_c*~marriage*gender*~age_cc*educ_c*~income_c	0.06	0.02	0.95
5 ~cult_c*~env_c*~soci_c*~eco_c*~marriage*~gender*age_cc*educ_c*~income_c	0.04	0.01	0.98
6 ~cult_c*env_c*~soci_c*~eco_c*~marriage*~gender*~age_cc*educ_c*income_c	0.04	0.01	0.99
7 ~cult_c*~env_c*soci_c*eco_c*~marriage*~gender*~age_cc*educ_c*income_c	0.05	0.01	0.95
8 ~cult_c*env_c*soci_c*eco_c*marriage*~gender*age_cc*~educ_c*~income_c	0.02	0.02	0.95
9 ~cult_c*env_c*soci_c*eco_c*~marriage*gender*~age_cc*educ_c*~income_c	0.06	0.02	0.92
10 ~cult_c*env_c*soci_c*~eco_c*marriage*~gender*~age_cc*educ_c*income_c	0.03	0.03	0.91
Solution coverage: 0.30; solution consistency: 0.94			

Table 5.21 Customers' brand attractiveness configurations indicating high local market frequency of visits

(A) **Findings for P6 frequent visits: customers' brand attractiveness configurations indicating high local market frequency of visits**			
Model: visit_c = f(loy_c, trus_c, satis_c, image_c); intermediate solution: frequency cutoff: 1.00; consistency cutoff: 0.80			
Model	**Raw coverage**	**Unique coverage**	**Consistency**
No models with high consistencies.			
(B) **Findings for P6 few visits: customers' brand attractiveness configurations indicating low local market frequency of visits**			
Model: not_visit_c = f(loy_c, trus_c, satis_c, image_c); intermediate solution: frequency cutoff: 1.00; consistency cutoff: 0.87			
Model	**Raw coverage**	**Unique coverage**	**Consistency**
1 ~loy_c	0.64	0.17	0.85
2 trus_c*~image_c	0.32	0.00	0.85
3 trus_c*~satis_c	0.37	0.00	0.90
4 satis_c*~image_c	0.28	0.00	0.86
5 ~trus_c*satis_c	0.34	0.01	0.91
6 ~satis_c*image_c	0.36	0.00	0.95
7 ~trus_c*image_c	0.36	0.00	0.94
Solution coverage: 0.76; solution consistency: 0.83			

visit frequency emerge with the addition of demographic dimensions into the brand attractiveness configurations. All three models include low income and all four dimensions of brand attractiveness, plus additional demographic terms. This pattern of findings is similar to the pattern found for sustainability and TLM visits.

In Table 5.22B, unmarried, young, and female are demographic terms relevant for 6 of the 12 models for negation of TLM visits, "relevant" in that female appears with the other two demographic terms for models 2, 3, 4, and 9, and models 1 and 5 include both genders (indicated by the absence of gender in these two models). Figure 5.2a shows that the demographic configuration (unmarried • young • female shoppers) alone indicates high (0.84) consistency of not TLM visit—however 4 of the 42 shoppers high in this demographic configuration are high in shopping visits. For the complete model 1 in Table 5.22B (i.e., eight terms include four negations for the four brand attractiveness dimensions), consistency is very high (0.98). Figure 5.2b contains these findings. (Consistency and coverage estimates vary slightly in Fig. 5.2b and model 1 Table 5.22B.)

The findings that unmarried • young • female shoppers along and in configurations for not visits and never in configurations of visit provide confirmatory evidence that these shoppers do not represent a favorable customer segment for nurturing TLM visits. Attempting to nurture these shoppers to visit the TLM frequently is less likely to be a successful retailing campaign than alternative strategies.

Table 5.22 Customers' brand attractiveness and demographic configurations indicating high versus low local market frequency of visits

(A) **Findings for P6x_C_frequent visits: customers' brand attractiveness and demographic configurations indicating high local market frequency of visits**

Model: visits_c = f(marriage, gender, age_c, educ_c, income_c, loy_c, trus_c, satis_c, image_c); intermediate solution: frequency cutoff: 1.00; consistency cutoff: 0.82

Model	**Raw coverage**	**Unique coverage**	**Consistency**
1 marriage*~gender*~age_cc*~educ_c*~income_c*loy_c*trus_c*satis_c*image_c	0.03	0.02	0.94
2 marriage*~gender*age_cc*~educ_c*income_c*loy_c*trus_c*satis_c*image_c	0.06	0.05	0.82
3 ~marriage*gender*age_cc*~educ_c*income_c*loy_c*trus_c*satis_c*image_c	0.05	0.05	0.86

Solution coverage: 0.13; solution consistency: 0.85

(B) **Findings for P6x_C_few visits: customers' brand attractiveness and demographic configurations indicating low local market frequency of visits**

Model: not_visit_c = f(marriage, gender, age_c, educ_c, income_c, loy_c, trus_c, satis_c, image_c); intermediate solution: frequency cutoff: 1.00; consistency cutoff: 0.90

Model	**Raw coverage**	**Unique coverage**	**Consistency**
1 ~marriage*~age_cc*educ_c*income_c*~loy_c*~trus_c*~satis_c*~image_c	0.14	0.04	0.97
2 ~marriage*~gender*~age_cc*educ_c*~income_c*~loy_c*~satis_c*image_c	0.07	0.02	0.98
3 ~marriage*~gender*~age_cc*educ_c*income_c*~loy_c*~trus_c*satis_c	0.07	0.01	0.99
4 ~marriage*~gender*~age_cc*educ_c*income_c*~loy_c*trus_c*image_c	0.06	0.01	1.00
5 ~marriage*~age_cc*educ_c*income_c*loy_c*trus_c*satis_c*image_c	0.11	0.03	0.92
6 ~marriage*gender*~age_cc*~educ_c*~income_c*~loy_c*~trus_c*~satis_c*~image_c	0.05	0.02	0.93
7 marriage*~gender*age_cc*~educ_c*~income_c*loy_c*trus_c*~satis_c*~image_c	0.02	0.01	0.96
8 marriage*~gender*age_cc*~educ_c*~income_c*~loy_c*trus_c*satis_c*~image_c	0.02	0.01	0.91
9 ~marriage*~gender*~age_cc*educ_c*~income_c*loy_c*~trus_c*satis_c*image_c	0.05	0.01	1.00
10 marriage*~gender*age_cc*educ_c*income_c*~loy_c*trus_c*~satis_c*~image_c	0.04	0.03	0.94
11 ~marriage*~gender*age_cc*educ_c*income_c*loy_c*trus_c*~satis_c*~image_c	0.04	0.01	1.00
12 ~marriage*gender*~age_cc*educ_c*~income_c*~loy_c*trus_c*satis_c*image_c	0.06	0.02	1.00

Solution coverage: 0.38; solution consistency: 0.94

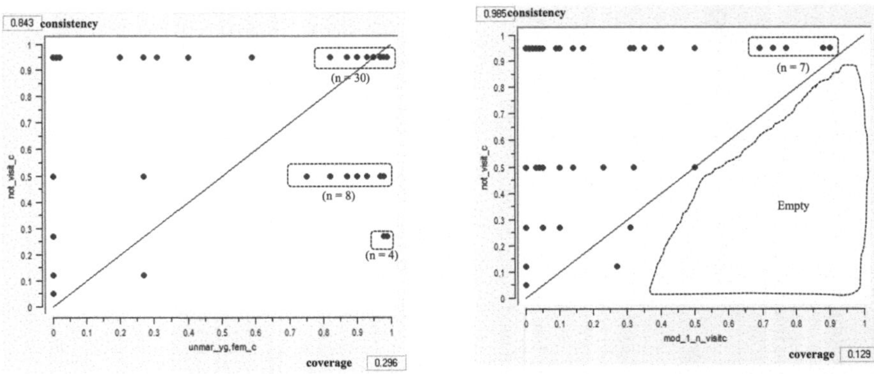

A. Unmarried • young • female shoppers ≤ ~Visit

B. Unmarried • young • female shoppers • ~image • ~loyal • ~ image • ~ satisfaction • ~ trust ≤ ~ Visit

Fig. 5.2 TLM not visit model for demographic configuration alone and same demographic configuration with attractiveness terms

18 Findings for P6: TLM Brand Attractiveness Configurations Without and with Demographics Indicating High and Low Expenditures

Tables 5.23 and 5.24 present the findings for TLM brand attractiveness without and with demographics, respectively, for indicating shoppers' expenditures. High consistency with prior findings for visits as the outcome is apparent from examining Tables 5.21 and 5.23 and Tables 5.22 and 5.24. TLM brand attractiveness configurations alone are unable to predict shoppers with high TLM expenditures (Table 5.23A). However, seven models have high consistencies for TLM brand attractiveness indicating shoppers having low expenditures (Table 5.23B). All seven of these models have at least one negative term for brand attractiveness. The findings here for low expenditures include the same models appearing in the findings for the negation of TLM visits. The coverage of these seven models is very high (0.74).

Table 5.24A includes four models of TLM brand attractiveness with demographic terms indicating shoppers with high TLM expenditures. Three of the four models include high scores for all four brand attractiveness terms along with demographic terms. Model 2 includes three negative and one positive (i.e., satisfaction) brand attractiveness terms along with demographic terms—a finding of negative brand attractiveness terms except for satisfaction indicating high expenditures is surprising—these shoppers are married, older, and males with high education and high income. A nuanced point here is that shoppers with high negative scores for one or more brand attractiveness attributes do not automatically indicate that they cannot be big spenders. 0 where brand attractiveness attributes alone may be unable to identify big spenders, such shoppers are likely identifiable by modeling both brand attractiveness attributes with demographic terms.

Table 5.23 Findings for P6 TLM Brand attractiveness indicating high and low expenditures

(A) **Findings for P6_C_high expenditures: customers' brand attractiveness configurations indicating high expenditures on local market**			
Model: exp_c = f(loy_c, trus_c, satis_c, image_c); intermediate solution: frequency cutoff: 1.00; consistency cutoff: 0.87			
Model	**Raw coverage**	**Unique coverage**	**Consistency**
No models with high consistencies.			
(B) **Findings for P6_C_low expenditures: customers' brand attractiveness configurations indicating low expenditures on local market**			
Model: not_exp_c = f(loy_c, trus_c, satis_c, image_c); intermediate solution: frequency cutoff: 1.00; consistency cutoff: 0.80			
Model	**Raw coverage**	**Unique coverage**	**Consistency**
1 ~loy_c	0.63	0.17	0.82
2 trus_c*~image_c	0.32	0.00	0.85
3 trus_c*~satis_c	0.37	0.01	0.89
4 satis_c*~image_c	0.27	0.00	0.80
5 ~trus_c*satis_c	0.30	0.01	0.80
6 ~satis_c*image_c	0.34	0.00	0.88
7 ~trus_c*image_c	0.33	0.00	0.83
Solution coverage: 0.74; solution consistency: 0.78			

Table 5.24B includes eight models of TLM brand attractiveness with demographic terms indicating shoppers with low TLM expenditures. Five of these eight models include the demographic terms of unmarried, female, and young—the XY plots for negation for attractiveness dimensions without and with demographics for indicating expenditures (now shown) are similar for the negative of visits (Fig. 5.2). Young, single, and male occur in two of the eight models. The warning is relevant here that young, single, male, or female represent parts of more complex models. However, similar to young, single, and female, the model (young•single•male ≤ ~expenditure) also has high consistency (0.89) and substantial coverage (0.28).

19 Findings for P7: Sustainability Dimensions and Brand Attractiveness Predicting TLM Visits and Not Visits

Using only on sustainability dimensions and the brand attractiveness conditions, Table 5.25 includes the configurations useful in predicting high TLM visit frequency and the configurations useful for predicting low TLM visit frequency None of the possible configurations were able to identify shoppers with high TLM visit frequency. However, Table 5.25 includes 18 models useful in identifying low visit frequency with all of these models having one more negation conditions for configurations that include 7 or all 8 of the sustainability and/or brand attractiveness conditions.

Table 5.24 Findings for P6 TLM Brand attractiveness with demographics indicating high and low expenditures

Model	Raw coverage	Unique coverage	Consistency
(A) **Findings for P6x_C_high expenditures: customers' brand attractiveness and demographic configurations indicating high expenditures on local market**			
Model: exp_c = f(marriage, gender, age_c, educ_c, income_c, loy_c, trus_c, satis_c, image_c); intermediate solution: frequency cutoff: 1.00; consistency cutoff: 0.83			
Model	**Raw coverage**	**Unique coverage**	**Consistency**
1 marriage∗~gender∗~age_cc∗~educ_c∗~income_c∗loy_c∗trus_c∗satis_c∗image_c	0.03	0.02	0.85
2 marriage∗gender∗age_cc∗educ_c∗income_c∗~loy_c∗~trus_c∗satis_c∗~image_c	0.05	0.05	0.83
3 marriage∗~gender∗age_cc∗~educ_c∗income_c∗loy_c∗trus_c∗satis_c∗image_c	0.07	0.06	0.96
4 ~marriage∗gender∗age_cc∗~educ_c∗income_c∗loy_c∗trus_c∗satis_c∗image_c	0.05	0.05	0.88
Solution coverage: 0.18; solution consistency: 0.88			
(B) **Findings for P7_C_few visits: customers' sustainability dimensions and brand attractiveness configurations indicating low local market frequency of visits**			
Model: not_visit_c = f(loy_c, trus_c, satis_c, image_c, cult_c, env_c, soci_c, eco_c); intermediate solution: frequency cutoff: 1.00; consistency cutoff: 0.92			
Model	**Raw coverage**	**Unique coverage**	**Consistency**
1 ~loy_c∗~trus_c∗~image_c∗~cult_c∗~env_c∗~soci_c∗~eco_c	0.21	0.03	0.90
2 ~loy_c∗satis_c∗~image_c∗~cult_c∗~env_c∗~soci_c∗~eco_c	0.16	0.01	0.92
3 ~loy_c∗~trus_c∗~satis_c∗image_c∗~cult_c∗env_c∗~soci_c	0.15	0.00	0.98
4 ~loy_c∗~trus_c∗~satis_c∗~cult_c∗env_c∗~soci_c∗eco_c	0.15	0.00	0.96
5 ~loy_c∗~satis_c∗~image_c∗cult_c∗env_c∗soci_c∗eco_c	0.20	0.02	0.96
6 loy_c∗trus_c∗satis_c∗image_c∗cult_c∗~soci_c∗~eco_c	0.13	0.01	0.89
7 loy_c∗trus_c∗satis_c∗image_c∗~cult_c∗env_c∗soci_c	0.19	0.03	0.92
8 loy_c∗~trus_c∗satis_c∗image_c∗cult_c∗env_c∗soci_c	0.18	0.01	0.96
9 loy_c∗~trus_c∗satis_c∗image_c∗cult_c∗env_c∗eco_c	0.17	−0.00	0.92
10 ~loy_c∗trus_c∗satis_c∗image_c∗env_c∗soci_c∗eco_c	0.20	0.00	0.93
11 trus_c∗~satis_c∗image_c∗cult_c∗env_c∗soci_c∗eco_c	0.21	0.02	0.97

12 ~loy_c*satis_c*image_c*cult_c*env_c*soci_c*eco_c	0.21	0.01	0.94
13 ~loy_c*trus_c*~satis_c*image_c*~cult_c*~env_c*~soci_c*~eco_c	0.13	0.00	0.97
14 ~loy_c*~trus_c*~satis_c*~image_c*cult_c*env_c*~soci_c*~eco_c	0.15	0.01	0.97
15 ~loy_c*~trus_c*~satis_c*~image_c*~cult_c*~env_c*soci_c*eco_c	0.15	0.01	0.97
16 loy_c*trus_c*~satis_c*~image_c*~cult_c*env_c*~soci_c*~eco_c	0.13	0.00	0.95
17 ~loy_c*trus_c*~satis_c*~image_c*cult_c*~env_c*~soci_c*eco_c	0.12	0.00	0.97
18 loy_c*trus_c*satis_c*image_c*cult_c*~env_c*soci_c*eco_c	0.14	0.01	0.96
Solution coverage: 0.56; solution consistency: 0.87			

Table 5.25 Sustainability and attractiveness configurations indicating visits and not visits

Model	Raw coverage	Unique coverage	Consistency
(A) **Findings for P7 Frequent visits: customers' sustainability dimensions and brand attractiveness configurations indicating high local market frequency of visits**			
Model: visit_c = f(loy_c, trus_c, satis_c, image_c, cult_c, env_c, soci_c, eco_c) ; intermediate solution: frequency cutoff: 1.00; consistency cutoff: 0.92			
Model	**Raw coverage**	**Unique coverage**	**Consistency**
No models with high consistencies.			
(B) Findings for P7_C_few visits: customers' sustainability dimensions and brand attractiveness configurations indicating low local market frequency of visits			
Model: not_visit_c = f(loy_c, trus_c, satis_c, image_c, cult_c, env_c, soci_c, eco_c) ; intermediate solution: frequency cutoff: 1.00; consistency cutoff: 0.92			
Model	**Raw coverage**	**Unique coverage**	**Consistency**
1 ~loy_c*~trus_c*~image_c*~cult_c*~env_c*~soci_c*~eco_c	0.21	0.03	0.90
2 ~loy_c*satis_c*~image_c*~cult_c*~env_c*~soci_c*~eco_c	0.16	0.01	0.92
3 ~loy_c*~trus_c*~satis_c*image_c*~cult_c*env_c*~soci_c	0.15	0.00	0.98
4 ~loy_c*~trus_c*~satis_c*~cult_c*env_c*~soci_c*eco_c	0.15	0.00	0.96
5 ~loy_c*~satis_c*~image_c*cult_c*env_c*soci_c*eco_c	0.20	0.02	0.96
6 loy_c*trus_c*satis_c*image_c*cult_c*~soci_c*~eco_c	0.13	0.01	0.89
7 loy_c*trus_c*satis_c*image_c*~cult_c*env_c*soci_c	0.19	0.03	0.92
8 loy_c*~trus_c*satis_c*image_c*cult_c*env_c*soci_c	0.18	0.01	0.96
9 loy_c*~trus_c*satis_c*image_c*cult_c*env_c*eco_c	0.17	-0.00	0.92
10 ~loy_c*trus_c*satis_c*image_c*env_c*soci_c*eco_c	0.20	0.00	0.93
11 trus_c*~satis_c*image_c*cult_c*env_c*soci_c*eco_c	0.21	0.02	0.97
12 ~loy_c*satis_c*image_c*cult_c*env_c*soci_c*eco_c	0.21	0.01	0.94
13 ~loy_c*trus_c*~satis_c*image_c*~cult_c*~env_c*~soci_c*~eco_c	0.13	0.00	0.97
14 ~loy_c*~trus_c*~satis_c*~image_c*cult_c*env_c*~soci_c*~eco_c	0.15	0.01	0.97
15 ~loy_c*~trus_c*~satis_c*~image_c*~cult_c*~env_c*soci_c*eco_c	0.15	0.01	0.97
16 loy_c*trus_c*~satis_c*~image_c*~cult_c*env_c*~soci_c*~eco_c	0.13	0.00	0.95
17 ~loy_c*trus_c*~satis_c*~image_c*cult_c*~env_c*~soci_c*eco_c	0.12	0.00	0.97
18 loy_c*trus_c*satis_c*image_c*cult_c*~env_c*soci_c*eco_c	0.14	0.01	0.96
Solution coverage: 0.56; solution consistency: 0.87			

The main conclusions for the findings in Table 5.25 include the following summary observations. One or more negation assessments in combination with positive attributes for sustainability and/or brand attractiveness are sufficient for identifying shoppers visiting the TLM infrequently but not frequently. Negative sustainability and brand attractiveness assessments alone can hold veto power over TLM visits, while positive sustainability and brand attractiveness assessments are insufficient for predicting high visit frequency. Thus, the TLM not engaging in sustainability and brand attractiveness building accomplishments may expect negation in shoppers' visit frequencies.

20 Findings for P7: Sustainability Dimensions and Brand Attractiveness Predicting High Versus Low Shopper TLM Expenditures

The findings in Table 5.26 confirm and extend the findings in Table 25. Table 5.26 includes asymmetric models for configurations of sustainability and brand attractiveness attributes predicting high as well as low TLM monthly expenditures. Table 5.26 includes no configurations of the eight dimensions of sustainability and brand attractiveness attributes that are able to predict high expenditures. Table 5.26 includes nine highly useful models in predicting low expenditures. All nine of these models include one or more negation attributes for sustainability and/or brand attractiveness.

The models in Table 5.26 for the negation of TLM expenditures are all complex; the models each include seven or all eight attributes for TLM sustainability and TLM brand attractiveness. Both positive and negative dimensions for all eight attributes appear among the models. Thus, focusing only on a symmetric main effect represents shallow reporting. Shoppers with low expenditures are identifiable with highly positive assessments for each of the eight sustainability and brand attractiveness dimensions for the TLM.

Table 5.27 illustrates the use of a rigorous "complex causal outcome" test for shoppers for the outcome of TLM low visits AND TLM low monthly expenditures. The analysis for Table 5.27 demonstrates the usefulness of testing complex causal outcomes. A shopper must have high scores in the negations for both visits and expenditures to have high scores in the configuration outcome in this test. Because they visit infrequently and spend little, such shoppers are the least desirable for the TLM to achieve high performance. The findings provide evidence that the 16 models in Table 5.27 are able to identify such shoppers. All 16 models include one or more negative attributes for sustainability and/or brand attractiveness.

Table 5.26 Customers' sustainability dimensions and brand attractiveness configurations indicating high expenditures in the TLM

(A) **Findings for P7 High expenditures: customers' sustainability dimensions and brand attractiveness configurations indicating high expenditures on local market**			
Model: exp_c = f(loy_c, trus_c, satis_c, image_c, cult_c, env_c, soci_c, eco_c); intermediate solution: frequency cutoff: 1.00; consistency cutoff: 0.90			
Model	**Raw coverage**	**Unique coverage**	**Consistency**
No models with high consistencies.			
(B) **Findings for P7 low expenditures: customers' sustainability dimensions and brand attractiveness configurations indicating low expenditures on local market**			
Model: not_exp_c = f(loy_c, trus_c, satis_c, image_c, cult_c, env_c, soci_c, eco_c); intermediate solution: frequency cutoff: 1.00; consistency cutoff: 0.91			
Model	**Raw coverage**	**Unique coverage**	**Consistency**
1 ~loy_c*~trus_c*~image_c*~cult_c*~env_c*~soci_c*~eco_c	0.22	0.07	0.91
2 ~loy_c*~trus_c*~satis_c*image_c*~cult_c*env_c*~soci_c	0.15	0.01	0.95
3 ~loy_c*~trus_c*~satis_c*~cult_c*env_c*~soci_c*eco_c	0.15	0.00	0.91
4 ~loy_c*~satis_c*~image_c*cult_c*env_c*soci_c*eco_c	0.19	0.02	0.92
5 ~loy_c*trus_c*~satis_c*cult_c*env_c*soci_c*eco_c	0.21	0.04	0.92
6 ~loy_c*trus_c*~satis_c*image_c*~cult_c*~env_c*~soci_c*~eco_c	0.13	0.00	0.94
7 ~loy_c*~trus_c*~satis_c*~image_c*~cult_c*~env_c*soci_c*eco_c	0.14	0.01	0.93
8 loy_c*trus_c*~satis_c*~image_c*~cult_c*env_c*~soci_c*~eco_c	0.12	0.01	0.92
9 ~loy_c*trus_c*~satis_c*~image_c*cult_c*~env_c*~soci_c*eco_c	0.12	0.00	0.94
Solution coverage: 0.40; solution consistency: 0.90			

Table 5.27 Complex causal outcome findings (shoppers with few_visits • low_expenditures)

Findings for P7 for complex causal outcome of few visits and low expenditures: customers' sustainability dimensions and brand attractiveness configurations indicating few visits and low expenditures on local market			
Model: n_vis_n_exp = f(loy_c, trus_c, satis_c, image_c, cult_c, env_c, soci_c, eco_c); frequency cutoff: 1.00; consistency cutoff: 0.91			
Model	**Raw coverage**	**Unique coverage**	**Consistency**
1 ~loy_c*~trus_c*~image_c*~cult_c*~env_c*~soci_c*~eco_c	0.21	0.03	0.83
2 ~loy_c*satis_c*~image_c*~cult_c*~env_c*~soci_c*~eco_c	0.16	0.01	0.83
3 ~loy_c*~trus_c*~satis_c*image_c*~cult_c*env_c*~soci_c	0.14	0.00	0.89
4 ~loy_c*~trus_c*~satis_c*~cult_c*env_c*~soci_c*eco_c	0.14	0.00	0.85
5 ~loy_c*~satis_c*~image_c*cult_c*env_c*soci_c*eco_c	0.19	0.01	0.86
6 ~loy_c*satis_c*image_c*cult_c*env_c*soci_c*eco_c	0.19	0.01	0.83
7 ~loy_c*trus_c*~satis_c*image_c*~cult_c*~env_c*~soci_c*~eco_c	0.12	0.01	0.87
8 ~loy_c*~trus_c*~satis_c*~image_c*cult_c*env_c*~soci_c*~eco_c	0.14	0.01	0.84
9 ~loy_c*~trus_c*~satis_c*~image_c*~cult_c*~env_c*soci_c*eco_c	0.14	0.01	0.85
10 loy_c*trus_c*~satis_c*~image_c*~cult_c*env_c*~soci_c*~eco_c	0.12	0.00	0.89
11 ~loy_c*trus_c*~satis_c*~image_c*cult_c*~env_c*~soci_c*eco_c	0.11	0.00	0.84
12 loy_c*trus_c*satis_c*image_c*~cult_c*env_c*soci_c*~eco_c	0.12	0.07	0.82
13 loy_c*~trus_c*satis_c*image_c*cult_c*env_c*soci_c*~eco_c	0.11	0.01	0.80
14 loy_c*trus_c*satis_c*image_c*cult_c*~env_c*soci_c*eco_c	0.13	0.01	0.81
15 ~loy_c*trus_c*~satis_c*cult_c*env_c*soci_c*eco_c	0.20	0.00	0.86
16 ~loy_c*trus_c*image_c*cult_c*env_c*soci_c*eco_c	0.22	0.00	0.84
Solution coverage: 0.49; solution consistency: 0.81			

21 Findings for the Deductive Models

This section presents the findings for the deductive models that the introduction and the theory section describe. The findings include details for testing for shoppers with high scores for each model for high visit frequency and high monthly expenditures.

21.1 Model 1 Findings: Age●Marital●Income●~Gender●Satisfy ≤ Visit

The computing with words for model 1 is the statement that high-income, older, married females who are satisfied with shopping in the TLM have high visit frequency (also, they have high monthly expenditures). The findings for model 1 include a low consistency index (0.541 for visit frequency), and the consistency index for monthly expenditures (0.713), though higher, is below the arbitrary cutoff set for the study (i.e., cutoff equal to 0.80). The XY plots for the two tests of the models appear as Fig. 5.3a, b. These findings and examination of the XY plots support the conclusion that the model is not useful for identifying shoppers with high membership scores for visit frequency nor for monthly expenditures.

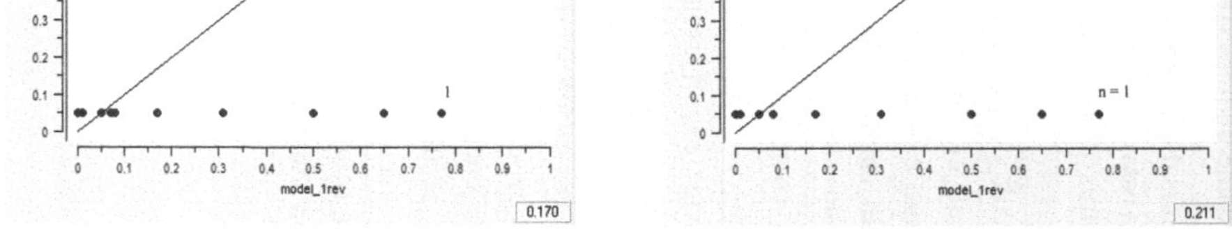

A. Age●Marital●Income●~Gender●Satisfy ≤ Visit

B. Age●Marital●Income●~Gender●Satisfy ≤ Expenditures

Fig. 5.3 Findings for model 1: Age•Marital•Income•~Gender•Satisfy ≤ Visit. Note: The usefulness of the model for expenditures is considerably greater than for visit frequency. However, the consistency for predicting shoppers having high scores for both models is below the arbitrary cutoff equal to 0.80. The model works poorly in identifying shoppers high in visit frequency and monthly expenditures. A total of seven shoppers have high scores (≥0.70) for the model for expenditures; six of the seven of these shoppers have high scores for monthly expenditures

21.2 *Findings for Model 2: A●~G●I●E●~M ≤ V*

The computing with words statement for model 2 is that older females with high income and high educations and not married have high visit frequency. Few shoppers have the configuration of demographic features that model 2 represents. Model 2 performs poorly in identifying shoppers with high TLM visit frequencies (consistency is equal to 0.55). Model 2 performs poorly in identifying shoppers with high monthly expenditures at the TLM (consistency equal to 0.47). This presentation of findings does not include the XY plots because the model performs poorly for identifying shoppers with high TLM visit frequency and high TLM monthly expenditures.

21.3 *Findings for Model 3: ~A•~G•I•~E•~M ≤ V*

The computing with words statement for model 3 includes the following attributes: young, female, high income, low education, and single engage in high visit frequency. The study includes no respondents with high scores for model 3. Consequently, model 3 works poorly in identifying shoppers with high membership scores for TLM visit frequency for TLM monthly expenditures. Presenting these findings does not include showing an XY plot for model 3.

21.4 *Findings for Model 4: A•G•~I•E•~M ≤ ~V*

The computing with words statement for model 4 includes the following attributes: older, male, low income, high education, and single shopper engage in low visit frequency. The study includes no respondents with high scores for model 4. Consequently, model 4 works poorly in identifying shoppers with high membership scores for TLM visit frequency for TLM monthly expenditures. Presenting these findings does not include showing an XY plot for model 4.

21.5 *Findings for Model 5: ~A•~G•~I•~E•M ≤ ~V*

Only one respondent had a high score in this model, and this respondent reported high, not infrequent, TLM visit frequency and had a modestly high TLM monthly expenditures. The model works poorly in identifying shoppers with TLM low visit frequency.

21.6 Findings for Model 6: Social•Economic•Culture•Environ mental ≤ Trust

The computing with words statement for model 6 is that shoppers with high TLM scores across all four sustainability dimensions have high scores on TLM trust. The findings support model 6—the consistency index for the model is 0.809. The odds of shoppers having high scores for TLM trust if they have high scores across all four sustainability dimensions is nearly 2 to 1. Figure 5.4 is an XY plot for the findings for model 6.

21.7 Findings for Model 7: Social•Economic•Culture ≤ Trust

The computing with words statement for model 7 is similar to the statement for model 6 without the inclusion of environmental sustainability. For model 7, the consistency index (0.791) is only slightly lower than the consistency index for model 6. Coverage for model 7 is slightly higher than coverage for model 6. Figure 5.5 is an XY plot for model 7. Together models 6 and 7 indicate a more parsimonious model which is sometimes feasible to use (model 7 versus model 6) without much loss in accuracy.

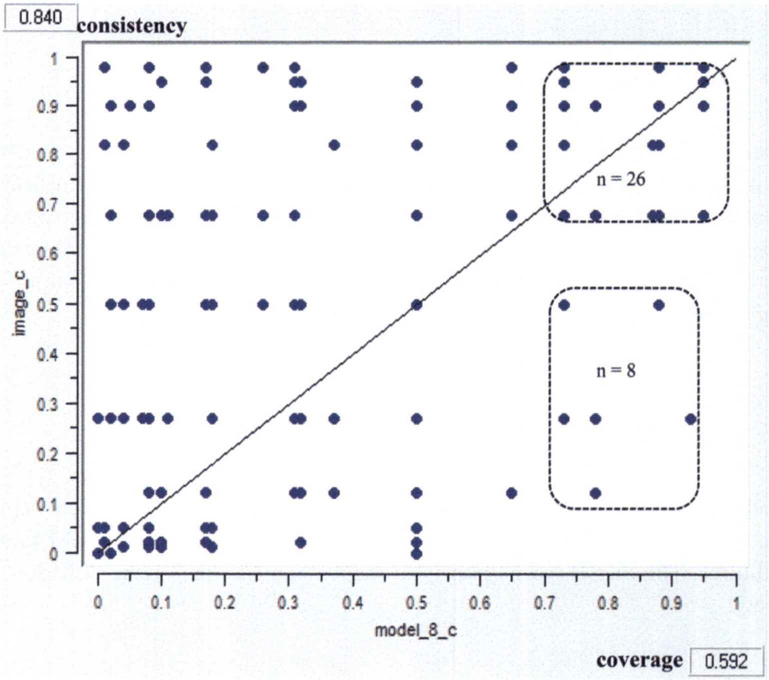

Fig. 5.4 Model 6 findings: Social•Economic•Culture•Environmental ≤ Trust

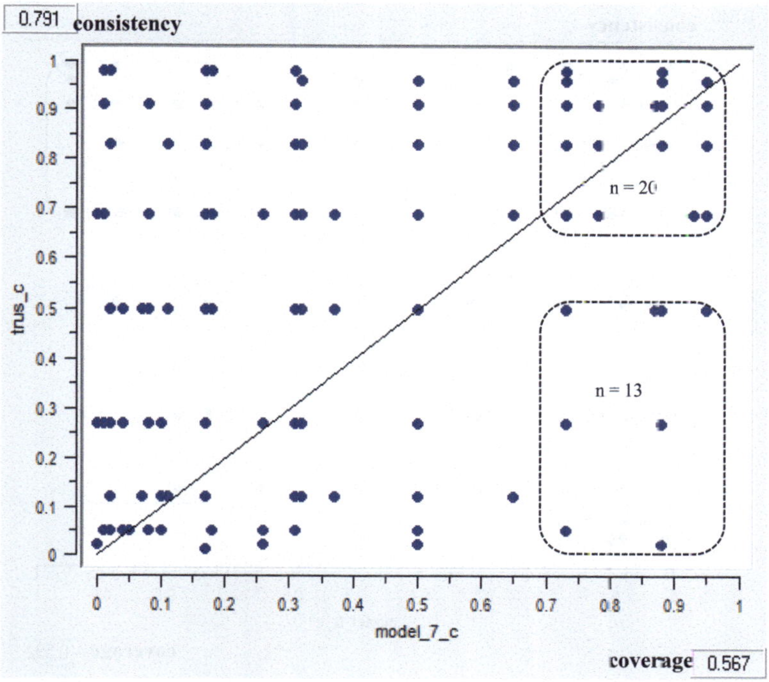

Fig. 5.5 Model 7 findings: Social•Economic•Culture ≤ Trust

21.8 Findings for Model 8: Social•Economic•Culture ≤ Image

Model 8 illustrates that the same fuzzy statement can be useful for different outcome conditions. The fuzzy statement for model 8 is the same as the fuzzy statement for model 7, but the outcome condition for model 8 is image rather than the outcome condition for model 7 (i.e., trust). Figure 5.6 presents the XY plot findings for model 8. Accuracy is somewhat higher for the findings for model 8 versus 7 (consistency for model 8 is equal to 0.84). For model 8 findings in Figure 5.6, the XY plot indicates that the odds are more than 3 to 1 for identifying shoppers with high scores for image among shoppers with high scores for the fuzzy statement for social•economic•culture sustainability. The substantial usefulness in some instance of the same model to identify shoppers having a high outcome condition is a useful observation from this analysis.

21.9 Findings for Model 9: ~Social•~Economic•~Culture•~Environmental ≤ ~Trust

The computing with words statement for model 9 is that shoppers with consistently high negation scores for each of the four TLM sustainability dimensions have negative trust toward the TLM. The findings for model 9 include high accuracy (consistency

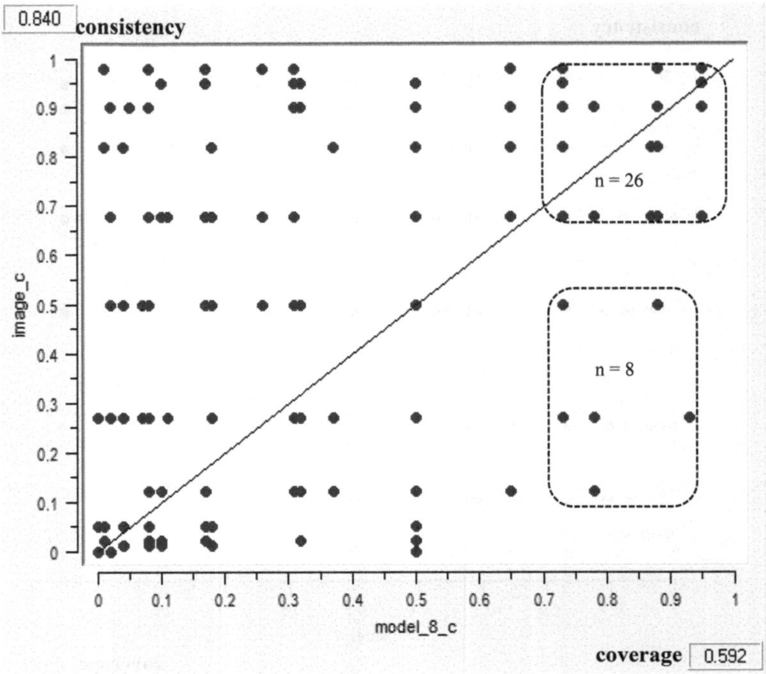

Fig. 5.6 Model 8: Social•Economic•Culture ≤ Image

is equal to 0.885) with the odds equal to 3 to 1 that shoppers with high scores on the fuzzy statement have high scores on the negation of trust outcome. These findings support the use of several indicators of negation—for in this instance—as hurdles for cases (i.e., shoppers) to confirm to predict cases for a specific outcome (negation of trust in this instance). Related to TLM sustainability, the findings confirm that most shoppers having consistently low sustainability assessments across different dimensions of sustainability have low trust in the TLM (Fig. 5.7).

21.10 Findings for Model 10: Social•Economic•Culture•Environmental ≤ Visits

Computing with words, model 10 states that shoppers with high scores across all four sustainability dimensions have high membership scores for visits. P5 expresses model 10. The findings do not support model 10. Conclusion: High membership scores in TLM sustainability are not a useful indicator for high scores in TLM visit frequency. However, this conclusion does not take into account the

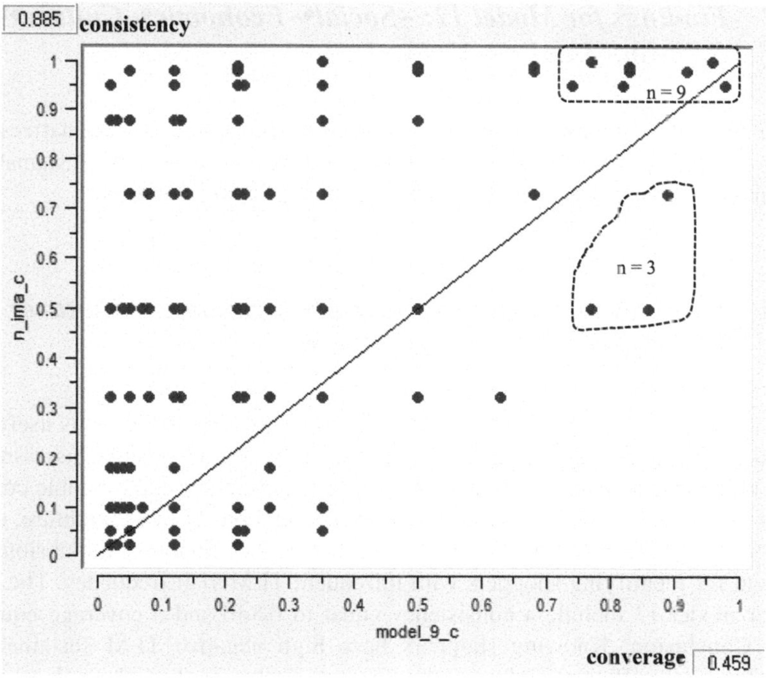

Fig. 5.7 Model 9: ~Social•~Economic•~Culture•~Environmental ≤ ~Trust

impact of shopper demographic contexts. Table 5.18 and its discussion do indicate high scores across all for TLM sustainability dimensions along with demographic information can discern shoppers with high TLM visit frequency. Context matters in assessing when shoppers' assessments of sustainability indicate TLM visit frequency accurately.

21.11 *Findings for Model 11: Social•Economic•Culture•Environmental ≤ Expenditures*

The findings for model 10 for TLM visit frequency apply for the findings for model 11 for TLM expenditures. Shoppers with high scores across the four sustainability dimensions do not exhibit high scores for TLM expenditures—however, this conclusion is modifiable by adding shoppers' demographic contexts into the TLM sustainability assessments. In the prior discussion for P5 and P5X, TLM sustainability assessments do indicate shoppers with high TLM expenditures for shoppers having certain demographic profiles. Table 5.20A describes these profiles.

21.12 *Findings for Model 12: ~Social•~Economic•~Culture•~Environmental ≤ ~Visits*

The findings do not support model 12. However, as Table 5.18 summarizes, the inclusion of shoppers' demographic configurations with negative TLM sustainability assessments is sufficient for indicating negative TLM visit frequency.

21.13 *Findings for Model 13: ~Social•~Economic•~Culture•~Environmental ≤ ~Expenditures*

The prior discussion and findings in Table 5.19 indicate that model 13 is useful for indicating shoppers visiting the TLM infrequently. While positive assessments across all four sustainability dimensions without considering demographic context are not useful for accurately predicting shoppers with high TLM expenditure, negative assessments for even one, two, or one of three sustainability dimensions are sufficient for identifying shoppers with infrequent TLM visit frequency. The findings for model 13 include a consistency equal to 0.840 and a coverage equal to 0.281. Conclusion: Knowing shoppers have high negative TLM sustainability dimensions is sufficient information to accurately predict their low TLM expenditures.

22 Discussion and Limitations

The present study confirms and deepens prior research (Ogle et al. 2004; Stone 1954; Wesley et al. 2006; Wu et al. 2014) that case-based reasoning, theory construction, and testing are useful for examining shoppers' behavior in retailing. The few available early (Stone 1954) and more recent (Wesley et al. 2006) case-based modeling studies on shopping outcomes support this general proposition that certain demographic identities influence retail shopping behavior. However, the modeling approaches used by Stone (1954), Ogle et al. (2004), and Wesley et al. (2006) do not include formal tests of specific case-based models. Heretofore, only a limited number of studies (e.g., Woodside et al. 2016; Wu et al. 2014) propose and test asymmetric, case-based, retailing, shopping, assessment, and decision models. The present study contributes by likely being the first to perform asymmetric modeling for predicting shoppers' visit frequencies and monthly expenditures. Such a modeling approach provides deep understanding, description, and prediction of shopper's TLM visit frequencies and monthly expenditures.

The findings show how to build in demographic information in asymmetric case models of mall shoppers' outcomes—and why doing so is very useful.

Heretofore, most studies attempting to model shoppers' assessments of retail stores apply symmetric models of variable relationships and exclude demographic information in constructing these models (e.g., Woodside and Trappey III 2001). The study's findings show that demographic conditions alone, sustainability conditions alone, and brand attractiveness conditions alone have mostly an inconsequential influence on accurate prediction of shoppers' TLM visit frequency. However, in configurations of demographics with sustainability dimensions or demographic with brand attractiveness dimensions, the study confirms the high value of modeling with demographics plus either of the two sets of additional conditions.

Examining the tables of findings confirms the perspective that TLM shopping visit frequency and monthly expenditures do not depend on individual demographic, sustainability, or brand attractiveness attributes but do depend on specific configurations of these attributes. Also, the same demographic, sustainability, and brand attractiveness attribute can serve to strength or weaken TLM visit frequency and monthly expenditures depending on the levels of the other attributes in the specific configurations. In comparison, focusing principally on reporting main effects of directional symmetric relationships provides shallow information even if the main effect is large. Rather than asking if a relationship is positive or negative and statistically significant, ask when does an antecedent condition positively and when does it negatively contribute in different combinations with other antecedent conditions to indicate cases having high membership scores in an outcome condition—and separately ask the same when questions for indicating cases having low membership scores for the same outcome condition.

The present study shows how and when sustainability attributes impact shoppers' TLM visit frequencies and monthly expenditures. All four sustainability attributes—social, economic, culture, and environment—are relevant in important way for many TLM shoppers. Negative TLM sustainability assessments have a profoundly negative impact on some shoppers' TLM visit frequencies and monthly expenditures. Managers of TLMs and managers of stores in TLMs should take heed/action to avoid negative assessments of the four sustainability attributes—even though the sustainability attributes alone are typically insufficient for identifying positive or negative shopping outcomes. With the inclusion of demographic conditions AND sustainability attributes, the case-based asymmetric models are able to identify sets of shoppers with positive TLM shopping outcomes and sets of shoppers with negative TLM shopping outcomes.

22.1 Limitations and Suggestions for Future Research

This study does not include tests for predictive validity using additional samples. Splitting the sample into two subsamples is possible, but the findings for testing a given asymmetric model always confirm the findings that the study reports—because the case-based method identifies all cases fitting the model. Using natural samples of shoppers for the same TLM or from different TLMs, shoppers not included in constructing and testing the models in the original study, is necessary to test for predictive validation. Future research on TLM shoppers' assessments should take this step. The models herein can serve as a source of models to test for their predictive validity and thus generality to additional TLMs and additional time periods. The present study includes a rather small (n = 162) sample size of shoppers. Future studies should attempt to include much larger numbers of TLM shoppers (e.g., $n \geq 1000$).

The study suffers from the use of one-shot shoppers' self-reports of TLM sustainability, brand attractiveness, visit frequencies, and monthly expenditures. Shoppers participating in study had little time for reflection and revisions of answers to the questions in the survey. The abilities of respondents to accurately report on their own assessments and behaviors are limited, as the iconic review by Nesbitt and Wilson (1977) illustrates from multiple studies. A possible method to measure and reduce bias would be to ask participants to provide their cell phone numbers and to telephone each a week after their completion of the first interview to ask them the same questions. An additional inducement could be provided to gain cooperation. If taking this step is too expensive to perform on the entire sample, applying this step for 10–20% of the first sample should be done. Performing this second interview on the same respondents would provide the respondents to think both explicitly and nonconsciously about their TLM assessments.

23 Contributions to Theory and TLM Management Practice

The present study provides new theory and tools useful for retailing researchers, store managers, and TLM managers seeking to understand how customer assessment store and/or mall performances in respect to sustainability practices and brand attractiveness. The study demonstrates how complexity theory tenets apply to the study of shoppers' assessments and models of their shopping behavior. The study presents evidence that researchers can have their cake and eat it too: They can construct models at the level of individual shoppers that are generalizable to include a number of shoppers and to additional samples of shoppers. The study is one of the first to demonstrate how and why retailing models of shoppers' assessments of malls (or stores) should include demographic attributes as integral components in the models.

The study concludes with a call to recognize the need to examine the antecedents and consequences of place sustainability and place expenditures as complex wholes—for the antecedent conditions, for measurement of sustainability and attractiveness, and for multiple outcomes. Relevant literature includes recognizing the telling weaknesses of symmetric modeling and findings (Hubbard 2016; Ziliak and McCloskey 2008); this recognition has had scant impact of the still dominant use of symmetric modeling and testing. Hopefully, the present study contributes to ending such practices by showing the relevancy of complexity theory and building/testing asymmetric models in retailing research.

References

Appleton, K., Beattie, R. Glover, A., Matthews, L., Orcutt, Pramis, A. J., & Shields, A. (2014). The World's most-visited tourist attractions. Travel & Leisure, November 10. Downloaded September 2, 2017 at http://www.travelandleisure.com/slideshows/worlds-most-visited-tourist-attractions#2, no pagination.

Arli, D. I., & Lasmono, H. K. (2010). Consumers' perception of corporate social responsibility in a developing country. *International Journal of Consumer Studies, 34*, 46–51.

Armstrong, J. S. (2012). Illusions in regression analysis. *International Journal of Forecasting, 68*, 689–694.

Ball, D., Simões Coelho, P., & Machás, A. (2004). The role of communication and trust in explaining customer loyalty: An extension to the ECSI model. *European Journal of Marketing, 38*, 1272–1293.

Bansal, P. (2005). Evolving sustainably: A longitudinal study of corporate sustainable development. *Strategic Management Journal, 26*, 197–218.

Baughn, C. C., & McIntosh, J. C. (2007). Corporate social and environmental responsibility in Asian countries and other geographical regions. *Corporate Social Responsibility and Environmental Management, 14*, 189–205.

Carroll, A. B. (1979). A three-dimensional conceptual model of corporate performance. *Academy of Management Review, 4*, 497–505.

Chen, Y. S. (2010). The drivers of green brand equity: Green brand image, green satisfaction, and green trust. *Journal of Business Ethics, 93*, 307–319.

Elkington, J. (1997). *Cannibals with forks: The triple bottom line of 21st century business*. Oxford: Capstone.

Faullant, R., Matzler, K., & Füller, J. (2008a). The impact of satisfaction and image on loyalty: The case of Alpine ski resorts. *Managing Service Quality: An International Journal, 18*, 163–178.

Faullant, R., Matzler, K., & Füller, J. (2008b). The impact of satisfaction and image on loyalty: The case of Alpine ski resorts. *Managing Service Quality: An International Journal, 18*, 163–178.

Fiss, P. C. (2007). A set-theoretic approach to organizational configurations. *Academy of Management Review, 32*, 1180–1198.

Fiss, P. C. (2011). Building better casual theories: A fuzzy set approach to typologies in organizational research. *Academy of Management Journal, 54*, 393–420.

Gigerenzer, G. (1991). From tools to theories: A heuristic of discovery in cognitive psychology. *Psychological Review, 98*, 254–267.

Hair, J. (2007). Knowledge creation in marketing: the role of predictive analytics. *European Business Review, 19*(4), 303–315.

Hubbard, R. (2016). *Corrupt research: The case for reconceptualizing empirical management and social science*. Thousand Oaks, CA: Sage.

Imagine your Korea (2017). Downloaded on September 26, 2017 at http://english.visitkorea.or.kr/enu/SHP/SH_EN_7_2.jsp?cid=2437620.

Jin, N., Lee, S., & Huffman, L. (2012a). Impact of restaurant experience on brand image and customer loyalty: Moderating role of dining motivation. *Journal of Travel & Tourism Marketing, 29*, 532–551.

Jin, N., Lee, S., & Huffman, L. (2012b). Impact of restaurant experience on brand image and customer loyalty: Moderating role of dining motivation. *Journal of Travel & Tourism Marketing, 29*, 532–551.

Jones, P., Comfort, D., & Hiller, D. (2011). Sustainability in the global shop window. *International Journal of Retail & Distribution Management, 33*, 882–892.

Jung, J. (2017). Shopkeeper interview #14, Seomun Market, Seomun market shopping study, Yonsei University, Yonsei Frontier Lab, March 7.

Kates, R. W., Clark, W. C., Corell, R., Hall, J. M., Jaeger, C. C., Lowe, I., McCarthy, J. J., Schellnhuber, H. J., Bolin, B., Dickson, N. M., Faucheux, S., Gallopin, G. C., Gruebler, A., Huntley, B., Jäger, J., Jodha, N. S., Kasperson, R. E., Mabogunje, A., Matson, P., Mooney, H., Moore, B., III, O'Riordan, T., & Svedin, U. (2001). Sustainability science. *Science, 292*, 641–642.

Kim, K. H., Jeon, B. J., Jung, H. S., Lu, W., & Jones, J. (2012a). Effective employment brand equity through sustainable competitive advantage, marketing strategy, and corporate image. *Journal of Business Research, 65*, 1612–1617.

Kim, K. H., Jeon, B. J., Jung, H. S., Lu, W., & Jones, J. (2012b). Effective employment brand equity through sustainable competitive advantage, marketing strategy, and corporate image. *Journal of Business Research, 65*, 1612–1617.

Kim, J., Taylor, C. R., Kim, K. H., & Lee, K. H. (2015a). Measures of perceived sustainability. *Journal of Global Scholars of Marketing Science, 25*, 182–193.

Kim, J., Taylor, C. R., Kim, K. H., & Lee, K. H. (2015b). Measures of perceived sustainability. *Journal of Global Scholars of Marketing Science, 25*, 182–193.

Maignan, I. (2001). Consumers' perceptions of corporate social responsibilities: A cross-cultural comparison. *Journal of Business Ethics, 30*, 57–72.

Maignan, I., & Ferrell, O. C. (2001). Corporate citizenship as a marketing instrument-concepts, evidence and research directions. *European Journal of Marketing, 35*(3/4).

McClelland, D. C. (1998). Identifying competencies with behavioral-event interviews. *Psychological Science, 9*, 331–3339.

McKenzie, S. (2004). *Social sustainability: Towards some definitions*, Hawke Research Institute Working Paper Series No. 27. University of South Australia: Adelaide.

Misangyi, V. F., Greckhamer, T., Furnari, S., Fiss, P. C., Crilly, D., & Aguilera, R. (2017). Embracing causal complexity: The emergence of a neo-configurational perspective. *Journal of Management*, forthcoming. Downloaded (September 9, 2017) and available at: http://journals.sagepub.com/doi/abs/10.1177/0149206316679252.

Nesbitt, R. E., & Wilson, T. D. (1977). Telling more than we can know: Verbal reports on mental processes. *Psychological Review, 84*, 231–259.

Nyadzayo, M. W., & Khajehzadeh, S. (2016). The antecedents of customer loyalty: A moderated mediation model of customer relationship management quality and brand image. *Journal of Retailing and Consumer Services, 30*, 262–270.

Ogle, J. P., Hyllegard, K. H., & Dunbar, D. H. (2004). Predicting patronage behaviors in a sustainable retail environment: Adding retail characteristics and consumer lifestyle orientation to the belief-attitude-behavior intention model. *Environment and Behavior, 36*, 717–741.

Oliver, R. I., Rust, R. T., & Varki, S. (1997). Customer delight: Foundations, findings, and managerial insight. *Journal of Retailing, 73*, 311–333.

Ordanini, A., Parasuraman, A., & Rubera, G. (2014). When the recipe is more important than the ingredients: A qualitative comparative analysis (QCA) of service innovation configurations. *Journal of Service Research, 17*, 134–149.

Park, S. (2012, August 22). Traditional markets epitomize Korean culture. *The Korea Times*. Retrieved 21 September 2017.

Pivato, S., Misani, N., & Tencati, A. (2008). The impact of corporate social responsibility on consumer trust: The case of organic food. *Business Ethics: A European Review, 17*, 3–12.

Ragin, C. C. (2008). Redesigning social inquiry: Fuzzy sets and beyond. Chicago: University of Chicago Press.

Rumelt, R. P. (2011). *Good strategy/bad strategy*. London, UK: Profile Books.

Sen, S., & Bhattacharya, C. B. (2001). Does doing good always lead to doing better? Consumer reactions to corporate social responsibility. *Journal of Marketing Research, 38*, 225–243.

Stone, G. P. (1954). City shoppers and urban identification: Observations on the social psychology of city life. *American Journal of Sociology, 60*, 36–45.

Trafimow, D. (2014). Editorial. *Basic and Applied Social Psychology, 36*(37), 1–2.

Van Herpen, E., Pennings, J. M., & Meulenberg, M. T. (2003). Consumers' evaluations of socially responsible activities in retailing (No. MWP-04). Lsg Marktkunde en Consumentengedrag.

Weick, K. E., & Sutcliffe, K. M. (2007). *Managing the unexpected* (2nd ed.). San Francisco: Jossey-Bass.

Wesley, S., LeHe, M., & Woodside, A. G. (2006). Consumer decision-making styles and mall shopping behavior: Building theory using exploratory data analysis and the comparative method. *Journal of Business Research, 59*, 535–548.

Wiese, A., Zielke, S., & Toporowski, W. (2015). Sustainability in retailing—Research streams and emerging trends. *International Journal of Retail & Distribution Management, 43*, 461–484.

Wilson, J. P. (2015). The triple bottom line: Undertaking an economic, social, and environmental retail sustainability strategy. *International Journal of Retail & Distribution Management, 43*, 432–447.

Woodside, A.G. (2014). Embrace•perform•model: Complexity theory, contrarian case analysis, and multiple realities, *Journal of Business Research 67*(12):2495–2503.

Woodside, A. G. (2016a). Embrace•perform•model: Complexity theory, contrarian case analysis, and multiple realities. *Journal of Business Research, 67*, 2495–2503.

Woodside, A. G. (2016b). Diffusion and adoption of good science: Overcoming the dominant logic of NHST and the reporting of rubbish. *Journal of Business-to-Business Marketing, 23*, 327–333.

Woodside, A. G. (2017). Releasing the death-grip of null hypothesis statistical testing (p < .05): Applying complexity theory and somewhat precise outcome testing (SPOT). *Journal of Global Scholars of Marketing Science, 27*, 1–15.

Woodside, A. G., & Trappey, R. J., III. (2001). Learning why some customers shop at less convenient stores. *Journal of Business Research, 54*, 151–159.

Woodside, A. G., Prentice, C., & Larsen, A. (2016). Revisiting problem gamblers' harsh gaze on casino services: Applying complexity theory to identify exceptional customers. *Psychology & Marketing, 32*, 65–77.

Woodside, A.G., Nagy, G., & Megehee, C. M. (2018). Four-corner outcomes in strategic management: Successful and unsuccessful paddling down versus upstream. In: Improving the Marriage of Modeling and Theory for Accurate Forecasts of Outcomes (Advances in Business Marketing and Purchasing, Vol. 25), Bingsely, UK: Emerald Publishing Limited, pp. 19–62.

Wu, P.-L., Yeh, S. S., Huan, T. C., & Woodside, A. G. (2014). Applying complexity theory to deepen service-dominant logic: Configural analysis of customer experience-and-outcome assessments of professional services for personal transformations. *Journal of Business Research, 67*, 1647–1670.

Zadeh, L.A. (1965). Fuzzy sets. *Information and Control*, 8, 38–353.

Zadeh, L. (1996). Fuzzy logic: Computing with words. *IEEE Transactions on Fuzzy Systems, 4*, 103–111.

Zadeh, L. (2010). Lofti Zadeh on CWW. In what computing with words means to me. In J. M. Mendel, L. A. Zadeh, E. Trillas, R. Yager, J. Lawry, H. Hagras, & S. Guadarrama (Eds.). *IEEE Computational Intelligence Magazine*, February, 20–26. https://people.eecs.berkeley.edu/~zadeh/papers/What%20Computing%20with%20Words%20Means%20to%20Me-CIM%202010.pdf. Accessed October 8, 2016.

Ziliak, S. T., & McCloskey, D. N. (2008). *The cult of statistical significance: How the standard error costs us jobs, justice, and lives*. Ann Arbor: University of Michigan Press.

Chapter 6
Cultures' Outcomes on Entrepreneurship, Innovation, and National Quality of Life

Arch G. Woodside, Carol M. Megehee, Lars Isaksson, and Graham Ferguson

Abstract This theoretical and empirical study applies complexity theory tenets to deepen understanding, explanation, and prediction of how configurations of national cultures and need motivations influence national entrepreneurial and innovation behavior and nations' quality of life (QOL). Also, the study examines whether or not high national ethical behavior is sufficient for indicating nations high in quality of life. Applying core tenets of complexity theory, the study constructs asymmetric, case-based (nations), explanation, and predictive models of cultures' consequences (via Schwartz's seven value dimensions) and implicit need motivations (via McClelland's three need motivations) indicating national entrepreneur and innovation activities and subsequent national quality of life and ethical behavior. The study includes testing configuration models empirically for predictive accuracy. The empirical examination is for a set of data for 24 nations in Asia, Europe, North and South America, and the South Pacific. The findings confirm the usefulness of applying complexity theory to learn how culture and motivation configurations support versus have negative consequences on nations' entrepreneurship, innovation, and human well-being. Nurturing of entrepreneur activities supports the nurturing of enterprise innovation activity, and their joint occurrence indicates nations achieving high quality of life. The findings advance the perspective that different sets of cultural value configurations indicate nations high versus low in entrepreneur and inno-

A. G. Woodside (✉)
Yonsei University, Yonsei Frontier Lab, Seoul, Republic of Korea
e-mail: arch.woodside@bc.edu

C. M. Megehee
Coastal Carolina University, Wall College of Business Administration, Conway, SC, USA
e-mail: cmegehee@coastal.edu

L. Isaksson
Queensland University of Technology, Building Z, School of Advertising, Marketing, and Public Relations, Brisbane, QLD, Australia
e-mail: lars.isaksson@qut.edu.au

G. Ferguson
Curtin University, School of Marketing, Bentley Campus, Perth, WA, Australia
e-mail: graham.ferguson@cbs.curtin.edu.au

A. G. Woodside (ed.), *Accurate Case Outcome Modeling*,
https://doi.org/10.1007/978-3-030-26818-3_6

vation activities. High entrepreneur activities without high innovation activity are insufficient for achieving high national quality of life. Achieving high ethical behavior supports high quality of life. This study is one of the first to apply complexity theory tenets in the field of entrepreneurship research. The study here advances the perspective that case-based asymmetric modeling of recipes is necessary to explain and predict entrepreneur activities and outcomes rather than examining whether or not variable relationships are statistically significant from zero.

Keywords Culture · Entrepreneurship · Innovation · Motivation · Nation · Quality of life

1 Introduction

This study presents a general theory of cultural value and motivation configurations influencing entrepreneurial behavior (EB) in nations, quality of life, and ethical behavior (i.e., the negation of a calibrated international measure of corruption). The study includes testing the model empirically. The general model and empirical examination improve on earlier research on the influence of cultural values on EB, QOL, and ethical behavior (Chelariu et al. 2008; Jin and Cho 2018; Woodside et al. 2016). By proposing and examining cultural value configurations as complex wholes rather than constructing and testing the relative sizes of impacts of individual values, this nascent asymmetric modeling approach to theory construction and empirical testing represents a true paradigm shift in the study of cultures' consequences on entrepreneurial behavior and quality of life.

Following this introduction, Sect. 2 describes cultures as complex wholes. Section 3 introduces the need motivations and entrepreneurship into the general theory. Section 4 provides a commentary on applications of complexity theory tenets into entrepreneurship research. Applying the complexity theory tenets, Sect. 5 updates the general theory of configuration models of culture and motives influencing entrepreneurial behavior and quality of life. Section 6 describes the details on the method of the study to test the general model. Section 7 presents the findings. Section 8 is a discussion of the study's contributions and limitations. Section 9 concludes with implications for national policy and training for nurturing entrepreneurial actions and high quality of life.

2 Culture as a Complex Whole

Appling Schwartz's seven cultural values, the present study examines the impacts of configurations of the seven in complex wholes on nations' entrepreneurship activity, quality of life, and ethical behavior (i.e., the negation of corruption). This perspective recognizes and goes beyond the theoretical possibility of 2187 configurations (i.e., 3^7) if data for each cultural value are discretized into terciles. Using terciles identifies when high (top 33%) versus low scores (bottom 33%) for a cultural value is useful in a complex whole for identifying an outcome; when neither are useful, the specific cultural value would not appear in the specific model. Discretizing in much finer gradations than terciles, available computer software (e.g., fuzzy-set qualitative comparative analysis (fsQCA)) will indicate models useful for a given outcome as well as models useful for indicating the negation of the outcome. The examples in the findings section provide additional description of the procedures.

Schwartz (2006, 2018) validates 3 cultural value dimensions containing 7 cultural value orientations based on his findings from data collected in 73 countries via 2 different instruments: the Schwartz Value Survey (SVS) and the Portrait Values Questionnaire (PVQ). The SVS data were collected from school teachers and college students in 67 nations in 1988 and again in 2000. The PVQ data were gathered from the European Social Survey (ESS) in 20 countries. Schwartz's seven cultural value orientations are summarized in three bipolar cultural value dimensions as follows (Schwartz 2018):

Autonomy versus embeddedness: The problem of defining the optimal relations and boundaries between the person and the group translates into the question: To what extent should people be treated as autonomous versus as embedded in their groups? "Autonomy" cultures treat people as autonomous, bounded entities. They encourage people to cultivate and express their own preferences, feelings, ideas, and abilities and to find meaning in their own uniqueness. There are two types of autonomy: Intellectual autonomy encourages individuals to pursue their own ideas and intellectual directions independently. Examples of important values in such cultures include broad-mindedness, curiosity, and creativity. Affective autonomy encourages individuals to pursue arousing, affectively positive personal experience. Important values include pleasure, exciting life, and varied life.

Egalitarianism versus hierarchy: The problem of ensuring coordination among people to produce goods and services in ways that preserve the social fabric translates into the question: How can human interdependencies be managed in a way that elicits coordinated, productive activity rather than disruptive behavior or withholding of effort? "Egalitarian" cultures urge people to recognize one another as moral equals who share basic interests as human beings. They socialize people to internalize a commitment to cooperate, to feel concern for the welfare of all, and to act voluntarily to benefit others. Important values in such cultures include equality, social justice, responsibility, help, and honesty. "Hierarchy" cultures rely on hierarchical systems of ascribed roles to ensure responsible, productive behavior. They define the unequal distribution of power, roles, and resources as legitimate and even desirable. People are socialized to take a hierarchical distribution of roles for granted, to comply with the obligations and rules attached to their roles, to show deference to superiors, and to expect deference from subordinates. Values of social power, authority, humility, and wealth are highly important in hierarchical cultures.

Harmony versus mastery: The problem of regulating the utilization of human and natural resources translates into the question: To what extent should individuals and groups control and change their social and natural environment versus leaving it undisturbed and unchanged? "Harmony" cultures emphasize fitting into rather than exploiting the social and natural world, accepting, preserving, and appreciating the way things are rather than trying to change them. Harmony cultures discourage efforts to bring about change and encourage maintaining smooth relations and avoiding conflict. Important values in harmony cultures include world at peace, unity with nature, protecting the environment, and accepting one's portion. "Mastery" cultures encourage active self-assertion by individuals or groups in order to master, direct, and change the natural and social environment and thereby to attain group or personal goals. They emphasize the desirability of active, pragmatic problem-solving that can produce "progress." Values such as ambition, success, daring, self-sufficiency, and competence are especially important in mastery cultures.

Schwartz (2006, 2018) stresses a theory specifying three bipolar dimensions of culture that represent alternative resolutions to each of three problems that confront all societies: autonomy versus embeddedness, egalitarianism versus hierarchy, and harmony versus mastery. "A societal emphasis on the cultural orientation at one pole of a dimension typically accompanies a de-emphasis on the polar type, with which it tends to conflict. For example, Russian culture tends to emphasize hierarchy but not the opposing orientation of egalitarianism. American and Israeli culture tend to emphasize mastery and to give little emphasis to harmony. The cultures of Iran and China emphasize hierarchy and embeddedness but not egalitarianism and intellectual autonomy" (Schwartz 2018, p. 9. The present study goes beyond Schwartz's polar opposites' perspective to examine the possibilities that specific cultures can and do conjoin values in result in seemingly "bipolar" stances. Thus, the possibility exists for a nation to emphasize both hierarchy and egalitarianism, while other cultures may emphasize both mastery and harmony—for accomplishing specific outcomes.

3 Need Motivations and Entrepreneurship

Within his "theory of needs," McClelland (1961/1976, 1965, 1987) along and with colleagues (McClelland et al. 1953; McClelland and Winter 1969; Miron and McClelland 1979) proposed a general theory of high "need for achievement (nAch)" supporting successful entrepreneurial start-ups, long-term firm success, and national well-being. His nAch theory includes the following propositions: first, national cultures influence specific nations' level (low versus high) of achievement motivation (nAch). Second, nations high in nAch frequently engage in entrepreneur start-ups more frequently per capita than nations low in nAch. Third, high versus low numbers of entrepreneur start-ups associate with enterprise success and high national wealth. Fourth, even though nAch is a long-term, national, culturally directed trait that changes over decades, if at all, individuals in small support groups are trainable to use tools that will increase their nAch. Fifth, individuals receiving training in nAch will engage more frequently in entrepreneur start-ups and gain greater individual wealth versus individuals not receiving such training. Remarkably for a psychology professor at Harvard (or elsewhere), McClelland founded a consulting firms (McBer & Co) and engaged in creating and providing training programs in nAch and measuring the long-term impact of individuals receiving the training. The findings for McClelland's studies support his theory at both the national (McClelland 1961) and individual/firm levels (McClelland and Winter 1969; Miron and McClelland 1979).

McClelland's theory of needs is an expansion on Murray's (1938) exposition on 20 universal human needs. McClelland (1961) focused on describing and measuring three types of people high in one of three core acquired motives: need for achievement (nAch), need for affiliation (nAff), or need for power (nPow). "Need for achievement (nAch)" is the motive to strive for success in particular situations in which a person's performance would be compared at against some type of standard. "Need for affiliation (nAff)" is the motive to strive for positive, sometimes intimate, personal relationships; affiliation can include concern over establishing, maintaining, or restoring a positive affective relationship with another person or persons. Need for power (nPow) is the motive to strive to control other people for one's own goals or to achieve higher goals for the greater good; people high in nPow seek neither recognition nor approval from others—only agreement and compliance.

While McClelland (1961) recognized the presence of all three motives in humans, his studies focus on examining the antecedents and outcomes of persons that he identified as possessing high levels of each of the three motives. His work focuses in particular on nations and individuals that he estimated to be high in nAch. Unlike the present study, McClelland's studies do not include examining the impact of national cultures on configuration of the three acquired motives. For example, he does not consider cultures' consequences on nations or individual persons high in all three motives (nAch•nAff•nPow) or nations having relatively low levels of nAch and nPow but a high level of nAff (~nAch•nAff•~nPow)—the USA is an example of the first configuration, and Denmark is an example of the second configuration. (The mid-level dot, "•", represents the Boolean algebra "AND" operation. The sideways tilde, "~", represents the negation of the score for the condition.) Based on the tenets of complexity theory (see below), the impacts of motives are unlikely to be predictable by identifying nations or individuals with high scores for one motive alone. The study of modeling for consistent "statistical sameness" impacts (Hubbard 2015) of cultures and motives on outcomes of nations and individuals requires examinations of configurations low and high on cultural and motive traits. The findings below support this approach to explaining when specific cultural configurations indicate high versus low entrepreneur behavior, high/low quality of life, and ethical/corrupt behavior.

The present study offers an independent test of McClelland's theory and applies core tenets of complexity theory to offer a more nuanced perspective of cultures' consequences on motives, entrepreneurial behavior, and quality of life. The study focuses on examining and enriching McClelland's theory at the macro (national) level; examining the effectiveness of McClelland's small-group training program to increase individuals' nAch levels is beyond the scope and purpose of the present study. The findings indicate the substantial benefits of embracing complexity theory for useful advances in theory construction on explaining the cultural antecedents indicating motive strength and the substantial ability of cultural and motive configurations to indicate low versus high entrepreneurial activity and low/high quality of life. The findings support the perspective that high nAch alone is neither necessary nor sufficient for high macro entrepreneurial activity or high quality of life. However, configurations of cultural and motive conditions that include high nAch do associate with high entrepreneurial activity and high quality of life.

Modeling using asymmetric screens (i.e., "computing with words," Zadeh 1996), the present study applies the complexity theory tenet that the causes for a negative outcome include unique antecedents to the causes for a positive or negative simple or complex outcome. Thus, asymmetric versus symmetric modeling is necessary to apply to gain deep understanding, explanation, and predictive accuracy of outcomes. Asymmetric modeling includes a different mind-set from symmetric modeling: while symmetric-based models (typically using multiple regression analysis (MRA) and structural equation modeling (SEM)) attempt to explain/predict both low and high directional relationships of independent variables and a single dependent variable, asymmetric models attempt to predict/explain high (and the negation of high) scores only for a simple or complex outcome. A specific asymmetric model is a statement that predicts cases with high scores for the antecedent statement will have high scores for an outcome condition—the asymmetric antecedent statement does not predict outcome scores for cases having low scores for the antecedent statement. "Computing [cases with high outcome scores] with words (CWW)" (Zadeh 1996) and constructing screening models are helpful tropes for clarifying asymmetric model construction. As the following method section elaborates, computing the negation of scores for an outcome permits the researcher to construct and test models of cases with high scores for negative outcomes (e.g., nations with low entrepreneurial actions and nations with low quality of life).

4 Complexity Theory

Applying complexity theory in marketing, management, and psychology represents more of an earthquake than a paradigm shift because complexity theory redirects theory construction and data analysis to the study of antecedents indicating outcomes rather than focusing on variables indicating relationships. Urry's (2005, p. 4) observation and related, additional, empirical evidence provide a starting place for describing and applying complexity theory: "Relationships between variables can be non-linear with abrupt switches occurring, so the same 'cause' can, in specific circumstances, produce different effects." Complexity theory includes the set of tenets appearing in Table 6.1.

The first tenet (T1): cases having high values for one antecedent condition (X) may be necessary, but a high X value is rarely sufficient for indicating cases only with high outcome values. For example, if high X represents males and low X represents females and high Y is player on the US National Football League (NFL), then X must be high for indicating high Y, but most high X cases (individuals) are

Table 6.1 Core Tenets (Ti) of complexity theory

Tenet	Concept	Description	Boolean expression
T1	Insufficiency	High X may be necessary, but this condition is insufficient for identifying high Y	X //→ Y
T2 T3 T4	Sufficiency Equifinality Contrarian	A recipe (configuration) may be sufficient for indicating high Y A few (e.g., X, R, T), not one, distinctly unique complex configurations of antecedent considered indicate the same outcome Both high X and low X associate with high Y Both high X and low X associate with low Y	X•S•T ≤ Y (X•R) ≤ Y + (~X•T) ≤ Y (X • R) ≤ Y + (~X • T) ≤ Y (X • W) ≤ ~Y + (~X • F) ≤ ~Y
T5	Causal asymmetry	Complex antecedent conditions for low Y are not the mirror opposite of complex antecedent conditions for high Y	(X • R) ≤ Y ≠ (~X • ~R) ≤ ~Y
T6	Emergence	System effects occurring in creating configurations of complex conditions are greater than the sum of the simple conditions (where nAch = high need for achievement, nAff = high need for affiliation, nPow = high need for power, and M = high average motives)	(nAch•nAff•nPow ≠ M)

Key: M, R, S, T, X, Y = antecedent and outcome conditions
"•" mid-level dot indicates the Boolean logical "AND" operation (conditions both left and right of the dot are present)
"~" sideways tilde indicates the negation of the condition (e.g., 1−X = ~X)

not NFL players. High X is a necessary but insufficient condition for predicting a high Y outcome. Thus, applying T1, high nAch alone is insufficient for indicating a high entrepreneurship nation consistently.

The second tenet: one or more configurations each with two or more antecedent conditions may accurately/consistently predict the outcome (high Y). Typically, three to eight conditions are sufficient for indicating an outcome of interest. Thus, one or more specific national or individual motive configurations may indicate cases high in entrepreneur behavior. For example, model 1 represents cases (i.e., nations or individuals) high in nAch, nAff, and nPow and the asymmetrical theoretical proposition that such cases high in all three motives are high in entrepreneur behavior (EB):

$$\text{nAch} \bullet \text{nAff} \bullet \text{nPow} \leq \text{EB} \tag{6.1}$$

The USA is representative of model 1. The USA has high scores on all three motives relative to other nations. The present study includes 2 among a sample of 24 nations that have high scores fitting the description of model 1: Canada and the USA. The findings section reports on the accuracy of model 1.

Equifinality is the third tenet (T3): a few, not just one, complex antecedent configurations indicate the same outcome asymmetrically. Model 2 versus model 1 serves as an example of T3. Model 2: nations or individuals high in nAff but low in nAch and low in nPow indicate high EB:

$$\sim \text{nAch} \bullet \text{nAff} \bullet \sim \text{nPow} \leq \text{EB} \tag{6.2}$$

The present study includes three nations fitting model 2's description: Denmark, the Netherlands, and Norway. The findings section reports on the accuracy of model 2. While software procedures are available for identifying several models identifying cases with high scores for a study's outcome, theory and insights based on prior empirical observation (i.e., abduction) should guide the possible identification of complex antecedent configurations that indicate the study's outcome. Armstrong's (2012) recommendation for directional, symmetrical, variable-relationship MRA to *not* rely on the stepwise operation in building/testing models also applies for outcome asymmetrical model building/testing.

T4: contrarian cases occur usually even when an XY relationship is highly significant using a null hypothesis significance test (NHST) (e.g., $r = 0.75$, $p < 0.001$) of a positive or negative main effect. Thus, for continuous variable data, discretizing by creating terciles or quintiles for both X and Y and cross-tabulating to form a 3-by-3 or 5-by-5 table results in cases appearing in all 9 or 25 cells in the table—almost always for large samples ($n \geq 100$) even when the effect size between the two variables is very high. Researchers typically recognize cases contrary to a hypothesis and findings as anomalies and usually ignored them in symmetric variable-relationship studies. However, researchers adopting a case-based, asymmetric, outcome perspective embrace Rumelt's (2011, p. 248) wisdom: "But to others, an anomaly marks an opportunity to learn something perhaps very valuable. In science, anomalies are the frontier, where the action is."

T5: theory and data analysis must account for causal asymmetry. T5 includes recognizing that the configurations indicating success, high performance, and additional Y outcomes include different ingredients (i.e., combinations of simple antecedent conditions and sometimes the same and different directionalities of the same conditions) than the configurations indicating failure, low performance, and additional Y outcomes. Consequently, the study of complex antecedent conditions indicating high entrepreneur behavior and high quality of life is unlikely to be informative for understanding and explaining low entrepreneur behavior and low quality of life. Weick (1987) presents this perspective in developing theory of the highly reliable organization (HRO) rather than attempting to advance theory of organizational financial success. HRO theory proposes thoughts, actions, and organizational mechanism for identifying and preventing undesirable procedures and outcomes. Entrepreneur and/or quality of life researchers embracing T5 will construct and test models of negative outcomes, for example, model 3:

$$\sim \text{nAch} \bullet \sim \text{nAff} \bullet \text{nPow} \leq \sim \text{EB}, \sim \text{QOL} \tag{6.3}$$

Model 3 includes nations with dictatorships and "extractive economies" (Omeje 2008) (e.g., Nicaragua, Nigeria, and North Korea) where governmental power is centralized in the hands of the few who seek to maintain power at all costs.

T6: system emergence effects reduce the number of cases meeting the minimum scores for all subdimensions of a multidimensional construct. Few nations are high on all three recognized metrics for entrepreneur behavior: culture and government policies in place that nurture entrepreneur start-ups (measured by the Global Entrepreneur Monitor (GEM) studies, Singer et al. 2014), replicative entrepreneur activity (measured by the Global Entrepreneurship and Development Index (GEDI), Ács et al. 2015), and innovative entrepreneur activity (measured the Global Innovation Index (GII), Dutta et al. 2015). Where variable-relationship focus studies typically would take an average standardized score across these three metrics to measure national EB, researchers adopting T6 would be more prone to identify nations high in EB that were high across all three metrics—reasoning the consistent indication of high QOL requires nations to be high on all three EB actions.

5 The Updated General Theory of Complex Wholes of Culture and Motives Influencing Entrepreneurial Behavior and Quality of Life

Figure 6.1 is a visual of the general theory of configurations of cultural values, motives, entrepreneur behavior, quality of life, and corruption. While additional theories of cultural values are available (e.g., Hofstede 2001; Inglehart and Baker 2000; Steenkamp 2001), Schwartz's (2006, 2018) cultural value theory was selected to examine cultural value configurations for the present study. Unlike Hofstede (2001), Inglehart and Baker's (2000), and Steenkamp's (2001) frameworks based on a posterior theorizing, Schwartz's cultural value dimensions are based on a priori theorizing (Schwartz 2006). Schwartz's approach views "cultural dimensions as forming an integrated, non-orthogonal system that distinguishes his interdependent

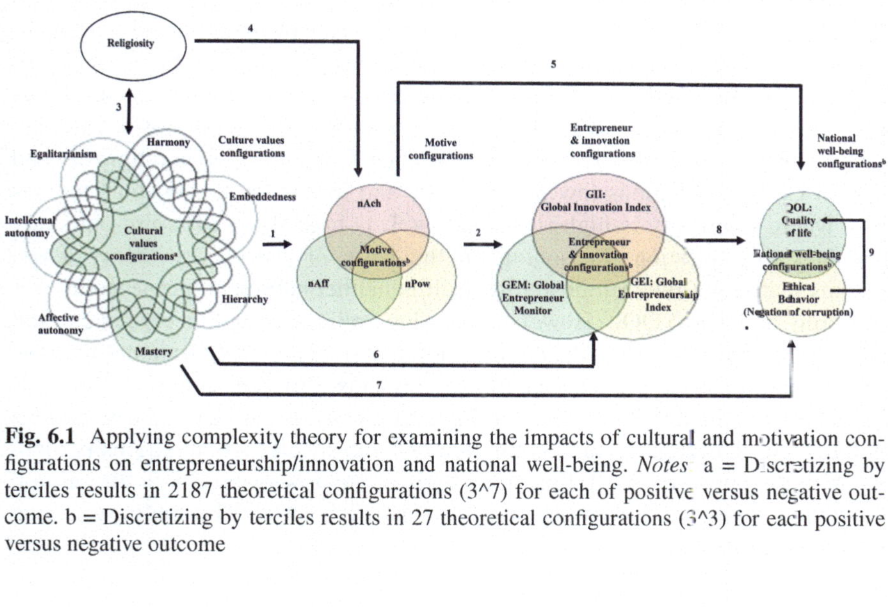

Fig. 6.1 Applying complexity theory for examining the impacts of cultural and motivation configurations on entrepreneurship/innovation and national well-being. *Notes* a = Discretizing by terciles results in 2187 theoretical configurations (3^7) for each of positive versus negative outcome. b = Discretizing by terciles results in 27 theoretical configurations (3^3) for each positive versus negative outcome

dimensions from Hofstede and Inglehart/Baker's orthogonal dimensions" (Schwartz 2006, p. 142). However, the developers of these four cultural value theories do not actually examine consequences of their theories from the perspective of culture as complex wholes—the ultimate definition of culture. Instead, the researchers apply symmetric examination of the directionality of variable relationships—a mismatch between theory and data analysis (cf. Fiss 2007, 2011).

Given that the lives of many residents in some nations appear to center on one religion (e.g., Christianity in the USA and Islam in Saudi Arabia), while most residents in other nations ignore religious practices (e.g., Denmark and Finland), the present study includes religiosity as a cultural trait. The present study embraces and examines the veracity of the following perspective by the Harvard Divinity School (2018), p. 1): "Religions are collections of ideas, practices, values, and stories that are all embedded in cultures and not separable from them. Just as religion cannot be understood in isolation from its cultural (including political) contexts, it is impossible to understand culture without considering its religious dimensions. In the same way that race, ethnicity, gender, sexuality, and socio-economic class are always factors in cultural interpretation and understanding, so too is religion."

Religiosity appears in Fig. 6.1 as an additional dimension of culture that associates with several configurations of cultural values. Using symmetric analysis to estimate the impact of religiosity independent of other influences on QOL of US communities and states, Ferriss (2000, 2002) finds and concludes "The religious community influences societal values. The greater the preponderance of religious adherents in the community the better the QOL" (Ferriss 2002, p. 210). Rather than asking if religiosity has a positive or negative directional relationship with QOL, research by Okulicz-Kozaryn (2010) indicates the usefulness of asking *when* high (low) religiosity is useful for indicating high (low) QOL: "This study shows that the relationship between religiosity and life satisfaction across nations is bimodal. Religious people tend to be either very satisfied or dissatisfied with life. The relationship between religiosity and life satisfaction is also two-dimensional" (Okulicz-Kozaryn 2010, p. 155). Unfortunately, Okulicz-Kozaryn (2010) applies only MRA and does not use configurations to report on nation case bases which nations are in each of the four corners: nations high in religiosity and high in QOL, nations high in religiosity and low in QOL, nations low in religiosity and high in QOL, and nations low in religiosity and low in QOL. Do some nations exist where religiosity is high (low) and QOL is high (low)? The present study examines this question. Reasoning on the basis of complexity theory tenets informs that high (low) religiosity by itself is neither sufficient nor necessary for indicating high (low) QOL. However, high (low) religiosity within certain cultural value configurations does indicate high (low) QOL. The configurational findings support both of these propositions. In the findings, different cultural value configurations that include low as well as high religiosity do, in fact, indicate nations high in QOL. Different cultural configurations that are all high in religiosity identify nations low in QOL. Religiosity—high versus low—appears in all cultural configuration models indicating both low and high QOL. The findings section includes specific details.

5.1 The First Proposition (P1): Cultural Value Configurations Indicate Motivational Needs

Arrow 1 in Fig. 6.1 expresses this first proposition. McClelland's (1961) theory of needs rests on the following foundational statement: cultures influence individuals' and nations' expressions for high nAch, nAff, and nPow. While McClelland (1961) does not address the corollary question, the proposition is worth stating and examining asymmetrically: cultures influence individuals' and nations' expressions for low nAch, nAff, and nPow. The findings section includes evidence supporting both statements: unique cultural configurations indicate the presence of high (low) membership scores for each of the three motivational needs. Different cultural configurations nurture different orientations toward the three basic needs.

5.2 P2: Need Configurations Indicate Entrepreneur Behavior (EB)

McClelland's (1961) perspective—high nAch supports high EB—runs counter to the complexity theory tenet that simple antecedent conditions are sufficient for indicating an outcome. Arrow 2 in Fig. 6.1 expresses proposition 2 that configurations of the three needs indicate nations with high EB, while additional configurations of the three needs indicate nations with low EB. While the findings below do not support McClelland's simple antecedent condition of nAch alone indicating high EB, the findings do support proposition 2 that a few, different configurations of the three needs indicate high as well as low EB.

5.3 P3): Religiosity Embeds in Cultural Values Configurations

Based on direct everyday observation as described earlier, religiosity varies considerably across national cultures. Rather than predicting that cultural values configurations indicate religiosity, the more useful perspective is that including religiosity with cultural values configurations enhances the ability for indicating the presence and absence of high (low) need motivations, EB, and QOL. The findings in the present study support this pervasive perspective.

5.4 P4: Religiosity Indicates Need Motivations Individually

While religiosity is a useful description of some cultures, religiosity alone is unlikely to be sufficient to indicate low or high nAch or nAff or nPow according to the tenets of complexity theory. Consequently, the findings should not support religiosity as a sufficient condition alone for the need motivations nor low/high EB nor low/high QOL. The findings do support this proposition from complexity theory. Thus, P4 is an inaccurate view of religiosity's singular impact on all three outcome configurations.

5.5 P5: Configurations of the Three Need Motivations Indicate Nations Low/High in QOL

Arrow 5 in Fig. 6.1 visualizes the proposition that need motivation configurations indicate nations high, as well as nations low, in QOL. While the individual need motivations do not indicate low or high QOL nations, the findings do support P5—different configurations of the three need motivations do indicate nations low, as well as high, in QOL. Arrow 5 in Fig. 6.1 serves to represent the impact of need motivation configurations on nations' ethical behavior. The study estimates whether or not configurations of the three needs indicate nations with high ethical behavior. Separately from predicting high ethical behavior, the study examines the ability of need configurations to predict nations high in corruption. The findings support all of the subparts of P5.

5.6 P6: Cultural Value Configurations/Religiosity Indicates Nations Low Versus High in EB

While focusing on directionality of relationships and not case-based configurations predicting specific outcomes, prior research includes the findings that embeddedness and hierarchy values associate with high entrepreneur activity in developing nations. In developed nations, the opposite values of autonomy and egalitarianism nurture high entrepreneurship (Fayolle et al. 2014; Liňán et al. 2013). Values serve to motivate behavior (Bardi and Schwartz 2003). Need motivations, in turn, are characterized by being more specific—psychologically closer to actual behavior (Fayolle et al. 2014; Schacter et al. 2011). Dodd and Gotsis (2007) provide an intensive review of religiosity and EB literature. They conclude that religiosity impact on EB is not a matter of positive or negative but religiosity associates with EB in complex ways. The theory for the present study includes the proposition that nations either low or high in religiosity are identifiable as being high in EB as well as nations low or high in religiosity are identifiable as being low in EB—when low versus high religiosity associates with low versus EB depends upon what additional antecedents occur in configurations that include low versus high EB. This view is the major takeaway from reading Dodd and Gotsis (2007). The present study identifies such configurations in some depth.

Arrow 6 in Fig. 6.1 visualizes the proposition that nations low versus high in EB are identifiable by different cultural value configurations that include cultural values and religiosity. The findings support P6 with and without the inclusion of religiosity. Thus, the conclusion receives support that different cultural values configurations serve to thwart, while additional cultural values configurations nurture, EB.

5.7 *P7: Cultural Value Configurations Indicate Nations Low Versus High in (a) QOL and (b) Corruption*

Arrow 8 in Fig. 6.1 visualizes the proposition that specific cultural value configurations are identifiable that indicate nations with low versus high QOL and low versus high corruption levels. The findings support P8. Religiosity is an ingredient in a few but not all the useful models indicating nations low versus high in QOL and corruption. The empirical examination includes assessing the four-corner configuration outcomes for nations high versus low in QOL by nations high versus low in corruption. The findings support the perspective that the configurations of cultural value configurations with and without religiosity do indicate nations high in the complex outcome condition, QOL•corruption. Such findings indicate the usefulness of examining not only complex antecedent conditions but complex outcome conditions as well. The findings include five nations identifiable to be high in both QOL and ethical (low corruption) behavior: Denmark, Finland, Germany, the Netherlands, and Switzerland.

5.8 *P8: Entrepreneur Behavior (EB) Configurations Indicate High/Low QOL and Corruption*

P8: High EB (i.e., model for GEM•GEDI•GII) indicate high QOL•Ethical (where "ethical" is the negation of corruption). The findings support P8 as well as indicate that the three separate EB metrics individually do not indicate this complex outcome. Consequently, national policy nurturing start-ups (GEM) without actually achieving start-ups (~GEDI) and without actually achieving high innovation behavior (~GII) does not achieve the desirable complex outcome (QOL•Ethical). Achieving the benefits of EB requires all pistons working—high GEM, high GEDI, and high GII. The findings section provides details.

5.9 *P9: High Ethical Behavior Indicates High QOL*

P9: High ethical behavior among nations is sufficient for indicating high QOL. P10 addresses the issue of the relevancy of ethical behavior for achieving high QOL. The study measures high ethical behavior as the negation of a national corruption index. The findings support viewing high ethical behavior as being a sufficient condition for achieving high QOL. This finding supports the perspective that the enactment and implementation of effective government laws and policies to reduce and eliminate the occurrence of corrupt behavior may be necessary steps for achieving high national QOL.

6 Method

The study uses secondary data from 8 original sources for 24 nations. Data are available for more than 24 nations from each original source, but data across all conditions in the general theory are available only for the 24 nations in the study. Appendices A and B include the raw data for the 24 nations.

6.1 Schwartz's Cultural Values Typologies

Schwartz's data for his seven "cultural value orientations" comes from survey research. He operationalized the value priorities of individuals with the Schwartz Value Survey that includes 56 or 57 value items (SVS: Schwartz 1992; Schwartz and Boehnke 2004). The SVS presents two lists of abstract value items (e.g., social justice, humility, creativity, social order, and ambition). The first list contains 30 items that describe potentially desirable end-states in noun form; the second contains 26 or 27 items that describe potentially desirable ways of acting in adjective form. An explanatory phrase in parentheses following the item further specifies its meaning. For example, "EQUALITY (equal opportunity for all)" or "PLEASURE (gratification of desires)" is a hedonism item. Respondents rate the importance of each value item "as a guiding principle in MY life" on a 9-point scale labeled 7 (of supreme importance); 6 (very important); 5, 4 (unlabeled); 3 (important); 2, 1 (unlabeled); 0 (not important); and 1 (opposed to my values). People view most values as varying from mildly to very important. This non-symmetrical scale is stretched at the upper end and condensed at the bottom in order to map the way people think about values, as revealed in pre-tests. The scale also enables respondents to report opposition to values that they try to avoid expressing or promoting. This is especially necessary for cross-cultural studies because people in one culture or subculture may reject values from other cultures.

"In order to use values in cross-cultural comparisons, their meanings must be reasonably similar across cultures. To assess which items have equivalent meanings across countries, separate multidimensional scaling analyses of the value items were run on gathered between 1988 and 2002 in each of 233 samples from 68 countries located on every inhabited continent (total $N = 67{,}145$). The samples included highly diverse geographic, cultural, linguistic, religious, age, gender, and occupational groups. Samples included those that represented a nation or a region in it, school teachers in grades k-12, undergraduate students from a variety of fields, adolescents, and adult convenience samples. The analyses revealed that 47 items showed reasonably equivalent cross-national meanings (Schwartz 2006; cf. Fontaine et al. 2008)" (Schwartz 2018, p. 12).

Schwartz (2018) provides the following brief summary for each of the seven cultural values types. "Autonomy" cultures treat people as autonomous, bounded entities. They encourage people to cultivate and express their own preferences, feelings, ideas, and abilities and to find meaning in their own uniqueness. There are two types of autonomy: *Intellectual* autonomy encourages individuals to pursue their own ideas and intellectual directions independently. Examples of important values in such cultures include broad-mindedness, curiosity, and creativity. *Affective* autonomy encourages individuals to pursue arousing, affectively positive personal experience. Important values include pleasure, exciting life, and varied life. "Embeddedness" cultures treat people as entities embedded in the collectivity. The expectation is that meaning in life follows largely through in-group social relationships, through identifying with the group, participating in its shared way of life, and striving toward its shared goals. Embedded cultures emphasize maintaining the status quo and restraining actions that might disrupt in-group solidarity or the traditional order. Important values in such cultures are social elicits coordinated, productive activity rather than disruptive behavior or withholding of effort. "Egalitarian" cultures urge people to recognize one another as moral equals who share basic interests as human beings. They socialize people to internalize a commitment to cooperate, to feel concern for the welfare of all, and to act voluntarily to benefit others. Important values in such cultures include equality, social justice, responsibility, help, and honesty. "Hierarchy" cultures rely on hierarchical systems of ascribed roles to ensure responsible, productive behavior. They define the unequal distribution of power, roles, and resources as legitimate and even desirable. People are socialized to take a hierarchical distribution of roles for granted, to comply with the obligations and rules attached to their roles, to show deference to superiors, and to expect deference from subordinates. Values of social power, authority, humility, and wealth are highly important in hierarchical cultures. "Harmony" cultures emphasize fitting into rather than exploiting the social and natural world, accepting, preserving, and appreciating the way things are rather than trying to change them. Harmony cultures discourage efforts to bring about change and encourage maintaining smooth relations and avoiding conflict. Important values in harmony cultures include world at peace, unity with nature, protecting the environment, and accepting one's portion.

In sum, the theory specifies three bipolar dimensions of culture that represent alternative resolutions to each of three problems that confront all societies: autonomy versus embeddedness, egalitarianism versus hierarchy, and harmony versus mastery. A societal emphasis on the cultural orientation at one pole of a dimension typically accompanies a de-emphasis on the polar type, with which it tends to conflict.

6.2 McClelland's Need Motivations

McClelland (1961) provides country values for the nAch, nAff, and nPow; he coded up to 21 stories for each of 40 countries from children's readers published between 1946 and 1955 for content reflecting the 3 needs. Coding instructions were based on extensive research examining the effects on thematic-apperception-test stories of experimental manipulations of the relevant needs. The validity of the resulting national motivation scores was demonstrated by their theoretically intelligible relationships to economic growth, birth rates, and political systems as McClelland (1961) reports.

6.3 Religiosity

The present study includes data from Stavrova et al.'s (2013) religiosity country norms. To operationalize the country's social norm of religiosity, Stavrova et al. (2013) constructed an indicator that reflects both descriptive and injunctive aspects of the norm. The descriptive aspect (national level of religiosity) was measured with a country's average values on the religiosity scale described above. The injunctive aspect of the norm (the degree to which religious beliefs are socially desirable and that those who do not share them are socially shunned) was captured by the following two questions. First, the respondents were given a list of qualities that children can be encouraged to learn at home (e.g., hard work, independence, and obedience) and were asked to select those that they consider especially important (up to five). One of these qualities was religious faith. We computed the percentage of respondents who mentioned religious faith in each country as an indicator of social desirability of religiosity. The second question reflected the degree of intolerance of non-religious behavior: "Politicians who don't believe in God are unfit for public service" (5-point strongly agree–strongly disagree scale). Each country's average score on this question was used as the second indicator of the injunctive aspect of the norm. Finally, each country's national level of religiosity, social desirability of religiosity, and intolerance of non-religious individuals were standardized and combined into an index of the country's social norm of religiosity (country-level Cronbach's alpha = 0.93). Higher values on this index indicate a higher average religiosity score in a country, a higher percentage of the population that believes that children must learn religious faith, and a higher degree of intolerance of non-religious people.

The study also uses data on national levels of religiosity made available by Finke and Adamcyzyk (2008). Finke and Adamcyzyk (2008) computed estimates of countries' religiosity by combining data from two studies: the 1991 International Social Survey Program (ISSP) and the World Value Surveys. For the ISSP, country-level estimates of religiosity were calculated by combining the individual religious importance and attendance measures. For the World Value Survey dataset, Finke and Adamcyzyk (2008) calculated the average country religiosity from the individual belief measure. For both surveys, higher numbers indicated higher country levels of religiosity. The findings of the studies by Stavrova, Fetchenhauer, and Schlösser (2013) and Finke and Adamcyzyk are highly similar. Appendix B includes the religiosity scores for all 24 nations as well as the data additional conditions in the study.

In addition, a third set of data on country religiosity was consulted. For use of a religiosity index in the present study makes use of country religious scores in WIN-Gallup (Global 2012, p. 14), "Religiosity and Atheism Index"; this index measures global self-perceptions on beliefs based on interviews with more than 50,000 adults across 57 countries in 5 continents. Share of residents who answer positive to the following question is used for the religiosity index. "Irrespective of whether you attend a place of worship or not, would you say you are a religious person, not a religious person, or a convinced atheist?" The percent share responding yes for Brazil was 86; China, 14; Germany, 51; Russia, 55; India, 79; the USA, 60; and Denmark, 18. The original shares responding yes were calibrated to a 0.00 to 1.00 scale using the logarithmic function in the fsQCA program to range from 0.00 to 1.00. The same calibration procedure was used for the religiosity estimates from the two other sources. The media calibrated score was selected for use in the present study.

6.4 Entrepreneur Behavior

The GEM 2014 data enable reporting on entrepreneur nurturing behavior (ENB) for 73 countries. GEM does not measure the number of entrepreneur start-ups per se; the evidence in the present study indicates the high scores on ENB in countries contribute to high scores on entrepreneur and innovative activities across countries. The following details on GEM's method appear in the GEM 2014 report (Singer et al. 2014). In order to capture the interactions between individuals and their environment, GEM uses two tools for collecting data on attributes, attitudes, and activities of individuals (Adult Population Survey) and on experts' opinions about the components of the "Entrepreneurial Conditions Framework (National Experts Survey)." A minimum of 2000 randomly selected adults (over 18 years old) must be surveyed in each country. The "Adult Population Survey" is conducted each year, from April to June, by independent survey vendors, using the GEM questionnaire. The "National Experts Survey" is conducted every year, during the same period of time, by GEM national teams, comprised of at least 36 experts (4 experts for each of the 9 components of the "Entrepreneurial Conditions Framework"), using the GEM questionnaire. The present study includes calibrated indexes of scores for seven ENB dimensions developed from the Global Entrepreneurship Monitor (GEM) data for 2014. Thus, the present study includes new, original data analysis from the GEM 2014 National Experts Survey. Bergmann et al. (2013) describe the use of data from the National Experts Survey for 109 scholarly studies. Bergmann et al.'s (2013) study of 109 studies is not intended to, and does not examine, the national cultures of the 73 nations as complex wholes or the experts' assessments of their nations' ENB as complex wholes as outcomes as well as antecedents to entrepreneur start-ups and innovations or the QOL in the nations. The present study does report on these complex-whole relationships. The following propositions express how cultural configurations impact ENB using the country-specific expert assessments in the GEM data as measures of ENB. P2.2a: Cultural configurations that favorably impact national ENB as measured by GEM are identifiable. P2.2b: Cultural configurations that negatively impact national ENB as measured by GEM are identifiable. P2.2c: Cultural configurations not indicative of positive or negative ENB as measured by GEM are identifiable—such cultural configurations are uninformative about ENB.

Ács et al. (2015) provide details for the computation of GEDI for 71 countries. While some overlaps of the dimensions of the three (GEM, GEDI, and GII) EB metrics are observable, the GEDI focuses on measuring replicative EB—the start-ups of new enterprises in a country which are mostly do not include technological innovations.

The present study uses the 2015 Global Innovation Index (GII) (Dutta et al. 2015) scores for the 28 nations in the study (the availability of data limited the number of countries in the study). Similar to the GEM and GEI (Global Entrepreneurship Index) scores, the GII (Global Innovation Index) national scores are composite scores. The GII scores focus on estimating innovation by firms in all industries and the implementation of national policies supporting innovative behaviors. The GII serves as a useful proxy for estimating Schumpeter's (1934, 1942) "creative destruction" innovative forms of entrepreneurship. The GII 2015 reports cover 141 economies around the world and use 79 indicators across a range of innovative policy and firm behavior themes. The GII scores overlap but are distinct in many ways from the GEM and GEI scores. In the GII 2015, Dutta et al. (2015), p. xvii, report that "Switzerland, the United Kingdom (UK), Sweden, The Netherlands, and the United States of America (USA) are the world's five most-innovative nations; at the same time, China, Malaysia, Viet Nam, India, Jordan, Kenya, Uganda, and a group of other countries are outpacing their economic peers in 2015." Cultural configurations that favorably impact national innovative entrepreneurship as measured by GII are identifiable. Other cultural configurations that negatively impact innovative entrepreneurship as measured by GII are identifiable.

6.5 *Quality of Life Metric*

The QOL metric in this study is a composite of countries' gross domestic product (GDP) adjusted for purchase price parity (PPP) and life longevity. QOL was calculated by first calibrating each nation's GDPppp and life longevity to membership scales ranging from 0.0 to 1.0. Second, the Boolean algebra operation to calculate the logical "AND" score was calculated for GDPppp•life_longevity—the lowest score across the two measures is the estimate for the configuration membership score for GDPppp•life_longevity. Country data for GDPppp and life longevity come from Woodside et al. (2016).

6.6 Ethical Behavior and Corruption Metric

The study included using country-level data in the "Corruption Perception Index" (Transparency International 2015). Ethical behavior was computed to be the negation of the calibrated country scores for corruption. For example, the calibrated corruption membership score for Mexico equals 0.98; thus, Mexico's ethical behavior score equals 0.02 (i.e., 1.00–0.98 = 0.02).

6.7 Analysis: Calibration Procedure

Calibrated membership scores are computed based on theory and knowledge of the distributions of variables for the data in the study. The fsQCA software requires the researcher to identify the membership score indicating the "threshold for full membership" equal to a 0.95 score, the membership score equal to 0.50 representing "maximum ambiguity," and the membership score of 0.05 representing the "threshold for full non-membership." Using these three scores, the software applies a logarithmic function to transform (i.e., calibrate) the raw scores into calibrated scores ranging from 0.0 to 1.00. Consequently, the resulting calibrated scores include no "statistical outliers"—very high scores and extremely high scores are equal to 1.00 calibrated scores, and very low scores and extremely low scores are equal to 0.00. Calibrated scores are not values but are membership qualifications of a condition.

Note that asymmetric analysis makes use of the term "condition," rather than the symmetric nomenclature for "variable." The calibration procedure is robust, and outcomes and interpretation of outcomes are usually very clear. For example, the findings and interpretation of these findings do not change substantively based on using fully non-membership scores covering 5 to 15 percent of the cases below the threshold, and the same perspective applies for full membership threshold scores equating to 5 or 10 percent above the threshold. For the present study, the value indicating that 10 percent of the country scores were higher was selected at the 0.95 membership score, the median value was selected as the 0.50 membership score, and the value indicating that 10 percent of the country scores were below was selected as the 0.05 membership score.

6.8 Computing with Words (CWW) for a Membership Score for a Configuration

Estimates of a membership configuration of complex antecedent or outcome conditions can be calculated for any combination of simple conditions. The configuration score is equal to the lowest calibrated score for all the conditions in the configuration. The configuration score represents the commonly that all the simple conditions share in the configuration. For example, consider the US calibrated scores for GEM (0.50), GEDI (0.99), and GII (0.92). The US calibrated, complex, configuration membership score for GEM•GEDI•GII equals 0.50.

6.9 Predicting the Impact of Antecedent Configuration Conditions on Outcomes

A consistency coefficient (C1) indicates the level of accuracy that high scores in an asymmetric configuration associate with high scores in a specific outcome of interest. The consistency index in asymmetric testing is analogous to a correlation ® in symmetric testing. However, unlike a correlation coefficient, low scores of a configuration do not influence a consistency coefficient since C1 is a measure of asymmetry and not symmetry. A coverage coefficient (C2) indicates the share of cases having high outcome scores applicable for the specific antecedent configuration. C2 is the asymmetric measure analogous to the coefficient of determination (r2) in symmetric analysis. However, unlike r2 estimates increasing when terms are adding to a regression model, C2 estimates decline as a researcher adds additional factors in an antecedent configuration because the researcher is adding restrictions to the applicability of a configuration by adding additional conditions. For the present study, the requirement was set that a model's consistency (C1)—for a complex antecedent configuration of antecedent conditions—needs to be equal or above 0.85 to conclude that the model has high consistency in predicting high scores for the focal outcome condition.

For the purposes of the present study, a conclusion was set ex ante that a model's coverage (C2) must be equal or above 0.20 for the model to be judged as having high coverage. Frequently, to achieve C1 $\geq$ 0.85, C2 is equal or less than 0.05; such findings are useful for identifying complex antecedent conditions indicating a few cases (e.g., countries) with high scores on the focal outcome condition. A researcher benefits from looking at the XY plot for such models that achieve high consistency whether or not coverage is high. In asymmetric modeling, high C1 is often achievable only when C2 is low; this perspective is another way of stating that contrarian cases occur in a set of data—the equifinality tenet in complexity theory that more than one path occurs for reaching the same outcome.

7 Findings

The findings include testing the specific models 1, 2, and 3 as well as exploratory testing of simple and complex antecedent and outcome configurations appearing in the general model (Fig. 6.1). The findings section first discusses the finding for P2—configurations of need motivations indicate entrepreneur behavior (EB).

7.1 *The US Need Motivation Model: Findings for Model 1—nAch•nAff•nPow ≤ EB*

The need motivation configuration, nAch•nAff•nPow, represents only two nations, Canada and the USA. Both Canada and the USA have high GEDI scores. Thus, the consistency (C1 = 0.88) is high for model 1 for GEDI, and coverage is modest (C2 = 0.26). For GEM and GII as outcome conditions, the consistency index for model 1 is below 0.85. Model 1 does not provide high consistency in estimating high scores for GEM and GII. Canada has a high score for GEM but not the USA. The USA has a high score for GII but not Canada. Figure 6.2 provides additional details. Note, for example, India achieves a high score in GEM but not in GEDI and GII.

7.2 *The "Danish Need Motivation Model": Findings for Model 2—~nAch• nAff •~nPow ≤ EB*

Model 2 is labeled the "Danish model." The configuration for this model is low nAch, high nAff, and low nPow—the Danish culture is highly neighborly. The findings for model 2 (~nAch• nAff •~nPow) include high consistencies for GEM (C1 = 0.87) and GII (C1 = 0.87). The consistency for GEDI is moderate. Figure 6.3 presents details for each of the three EB for model 2. Notice that the Danish model is also the "Netherlands model" and the "Norway model"—all three nations have high membership scores on model 2. However, Norway's membership scores in GEDI and GII are not high. Norway's governmental laws and policies encourages entrepreneurship, but this encouragement associates less with replicative and innovative entrepreneur behavior than in Denmark and the Netherlands.

7.3 *Findings for the German Need Motivations' Configuration Model*

For comparison purposes, this section describes findings for an additional nation need motivation model with Germany as the focal nation. Germany has nAch and nAff scores above 0.50 and a nAff score below 0.50. Thus, a computation was made of this "Germany need motivations' configuration" for each country using nAch•~nAff•nPow. (The "Germany need motivations' configuration" not only applies to Germany but also to additional nations (e.g., India and Spain)). Figure 6.4 presents the findings for the need motivations' model representing high scores for Germany, India, and Spain.

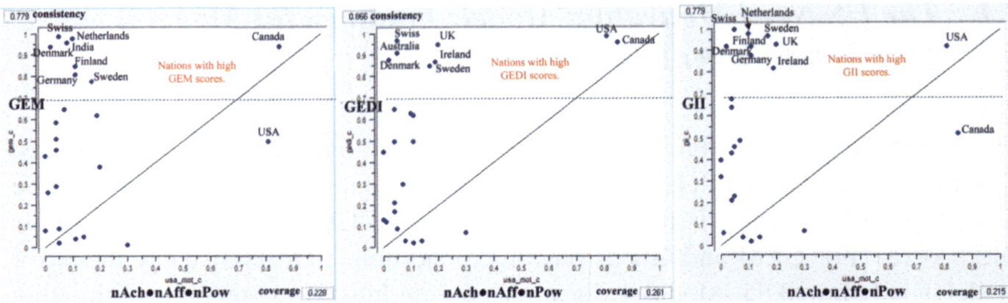

Fig. 6.2 USA's motives model (nAch•nAff•nPow) impact on GEM, GEI, and GII. *Notes*: (**a**) GEM (Global Entrepreneur Monitor) provides national indexes assessing nations' entrepreneurial nurturing behavior (ENB) across seven dimensions: 1, sufficient venture capitalist funding; 2, national government policy supports new and growing firms; 3, primary and secondary education encourages creativity and initiative; 4, sufficient suppliers, subcontractors, and consultants to support start-ups; 5, new and growing firms can easily enter new markets; 6, national culture highly supportive of individual success; 7, physical infrastructure provides good support for start-ups. (**b**) GEDI (Global Entrepreneur Index) provides country-level composite estimates of individual firm data on business start-ups that mostly represent replicative entrepreneurial activity; the GEI focuses more on entrepreneurial acts versus GEM's focus on ENB. (**c**) GII (Global Innovation Index) provides national composite scores estimating innovation by firms in all industries and the implementation of national policies supporting innovative behaviors. The GII serves as a useful proxy for estimating Schumpeter's (1942) "creative destruction" innovative forms of entrepreneurship. Findings: Adopting the rule that a consistency index equal or higher than 0.85 indicates a useful model for indicating high scores for an outcome, the USA motives model performs well only for indicating GEDI (replicative entrepreneurship). Additional motives models other than the USA motive configuration model are necessary for indicating countries high in GEM, GEDI, and GII

The findings indicate that the model is not useful for identifying nations high in GEM, GEDI, and GII scores consistently. The findings for the three national models (i.e., Denmark, USA, and Germany) indicate greater impact of the Danish model (i.e., ~nAch•nAff•~nPow) for indicating nations with high scores on GEM and GII with the US model (i.e., nAch nAff•nPow) having the greater impact for indicating nations high in GEDI (i.e., replicating entrepreneur activity). Note the Denmark need motivations' model is also the same for the Netherlands. Both these nations have national policies seeking to nurture high standards of social welfare processes and outcomes. Such policies as the following may connect positively to the development of EB, especially GEM and GII activities.

> What makes life so worth living in Denmark are the country's strong social policies, say experts. From cradle to grave, the Danish state knows how to help its people: maternity or paternity leave—parents can decide to split it up—is a full year. Nurseries and then schools are both guaranteed and paid for by the municipality. University education is tuition free and students can apply for living expenses. Families with children receive a generous allowance, regardless of income level. Even elderly care, such as senior centers, adapted transport and meals-on-wheels programs are paid for the by the state. (Schuetze 2018)

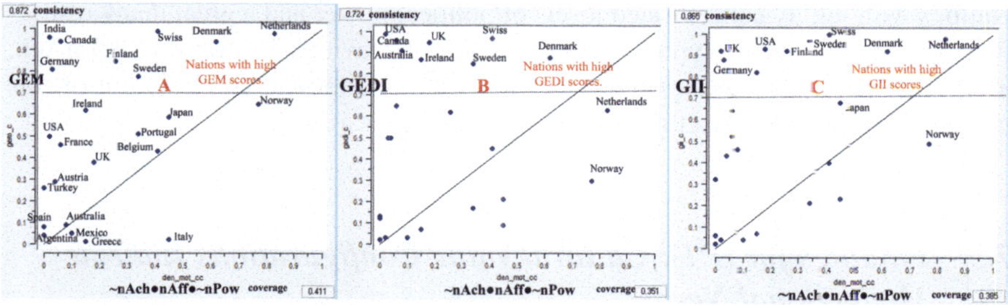

Fig. 6.3 Denmark motives model (~nAch•nAff•~nPow) impacts on GEM, GEI, and GII. *Notes*: (**a**) GEM (Global Entrepreneur Monitor) provides national indexes assessing nations' entrepreneurial nurturing behavior (ENB) across seven dimensions: 1, sufficient venture capitalist funding; 2, national government policy supports new and growing firms; 3, primary and secondary education encourages creativity and initiative; 4, sufficient suppliers, subcontractors, and consultants to support start-ups; 5, new and growing firms can easily enter new markets; 6, national culture highly supportive of individual success; 7, physical infrastructure provides good support for start-ups. (**b**) GEDI (Global Entrepreneur Index) provides country-level composite estimates of individual firm data on business start-ups that mostly represent replicative entrepreneurial activity; the GEI focuses more on entrepreneurial acts versus GEM's focus on ENB. (**c**) GII (Global Innovation Index) provides national composite scores estimating innovation by firms in all industries and the implementation of national policies supporting innovative behaviors. The GII serves as a useful proxy for estimating Schumpeter's (1942) "creative destruction" innovative forms of entrepreneurship. *Findings*: The motives' configuration for Denmark (~nAch•nAff•~nPow) indicates high GEM (A) membership and GII (C) membership scores consistently, but the XY plot does not indicate that the Danish motives model supports high GEDI scores. The findings are asymmetric: additional models of motive are necessary to report for other nations having high GEM, GEDI, and GII membership scores for the nations low on the Danish motives model

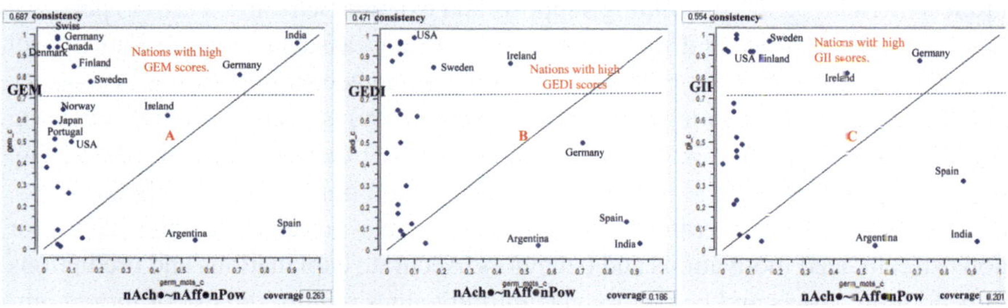

Fig. 6.4 Germany's motives model (nAch•~nAff•nPow) impact on GEM, GEI, and GII. *Findings*: The consistencies are low (<0.80) for all three outcome outcomes. The German motives model alone does not indicate high predictive accuracy for high GEM, GEDI, or GII scores

7.4 *The Findings for Model 3: ~nAch•~nAff•nPow ≤ ~EB, ~QOL*

The study includes no country representing the need motivation configuration for model 3. Consequently, this findings section is able to examine model 3 only as the thought experiment in the prior discussion on Nicaragua, Nigeria, and North Korea.

Nations with highly concentrated levels of political power and a great many restrictions on day-to-day life typically do not exhibit behaviors illustrating high nAch, nAff, and EB. These nations have very low QOL. Empirical confirmation of this perspective would be a useful step once data becomes available.

7.5 *Findings for (P1): Cultural Value Configurations Indicate Motivational Needs*

The findings support P1: different cultural value configurations indicate high (low) nAch, nAff, and nPow. Details appear in Tables 6.2a, 6.2b, and 6.2c, for the three need motivations, respectively. Additional details appear in Appendices D, E, and F for the inclusion of religiosity with cultural value configurations for indicating each of the three need motivations. Mastery and hierarchy cultural values are high for all three models indicating high nAch in Table 6.2a. The other five cultural values vary in their positive and negative inclusions among the three models. The findings support the usefulness of seven distinct cultural values—all seven appear in the three models for nAch. Given the focus of high mastery on values such as ambition, success, daring, self-sufficiency, and competence, the finding that mastery is an ingredient in all three models indicating high nAch is unsurprising and provides nomological validity of high mastery as an indicator for high nAch. However, high mastery alone is insufficient for indicating nations high in nAch (i.e., the consistency of mastery alone and nAch is equal to 0.68).

Note that both mastery and the negation of harmony are ingredients in all three models in Table 6.2a. This finding supports and extends Schwartz's (2018) perspective that high mastery and the negation of harmony associate together to impact motivations and behavior. However, the complex antecedent condition of mastery and the negation of harmony (i.e., mastery•~harmony) is insufficient for indicating high nAch (consistency = 0.74). Schwartz's (2018) over focuses on identifying nations by one of three difference scores (i.e., autonomy minus embeddedness, egalitarianism minus hierarchy, and harmony minus mastery) as antecedents for downstream need motivations and behaviors. Accurate explanations and predictions of need motivations and behaviors via cultural values require the use of more complex cultural value configurations than using difference scores. For example, hierarchy appears in all three models in Table 6.2a for nAch. Including hierarchy to construct mastery•~harmony•hierarchy results in a consistency in predicting high nAch equal to 0.79—a consistency higher than mastery alone or mastery•~harmony but well below the consistencies of the models in Table 6.2a for nAch. Thus, cultural value recipes serve to explain and predict individual need motivations, but these recipes are more complex than using configurations of two to six simple conditions. Schwartz's (Schwartz 2018) theory is remarkably useful given explanations and predictions that include configurations of all seven value dimensions.

Table 6.2a For P1: Cultural value configurations indicating high nAch and the negation of nAch

(A) **Cultural values configurations accurately indicating high nAch**			
Model: nAch_c = f(egalitar_c, intel_auto_c, aff_auto_c, mast_c, hier_c. embed_c, harm_c)			
INTERMEDIATE SOLUTION: frequency cutoff: 1.00; consistency cutoff: 0.94			
Models	**Raw coverage**	**Unique coverage**	**Consistency**
1. egalitar_c•~intel_auto_c•aff_auto_c•mast_c•hier_c•~embed_c•~harm_c	0.15	0.06	0.99
2. ~egalitar_c•~intel_auto_c•aff_auto_c•mast_c•hier_c•embed_c•~harm_c	0.11	0.04	1.00
3. egalitar•~aff_auto_c•mast_c•hier_c•~embed_c•harm_c	0.11	0.06	0.94
Solution coverage: 0.25; solution consistency: 0.97			
(B) **Cultural values configuration accurately indicating negation of nAch**			
Model: not_nAch_c = f(egalitar_c, intel_auto_c, aff_auto_c, mast_c, hier_c, embed_c, harm_c)			
INTERMEDIATE SOLUTION: frequency cutoff: 1.00; consistency cutoff: 0.87			
Model	**Raw coverage**	**Unique coverage**	**Consistency**
1. egalitar_c•intel_auto_c•~aff_auto_c•~mast_c•~hier_c•harm_c	0.17	0.17	0.89
Solution coverage: 0.17; solution consistency: 0.89			

Six cultural values appear in the negation model for nAch (i.e., ~nAch) at the bottom of Table 6.2a. Note that mastery and hierarchy are both negative in the cultural values model indicating ~Ach. At first blush, mastery and hierarchy appear to have symmetrical positive associations with nAch. However, this view is incorrect: neither mastery nor hierarchy by themselves has symmetric positive associations with nAch. The complex configurations of cultural value models indicate nations for high nAch is the accurate perspective to glean from Table 6.2a.

Additional findings for cultural values configurations and need motivations appear in Table 6.2b for nAff and nPow, respectively. These models may serve as initial benchmarks for identifying how cultural values configurations indicate the two basic need motivations. The main focus here is the finding that the cultural values configurations indicate nations with high (low) membership scores for each of the three need motivations is evidence that supports the general model of cultural values indicating need motivations. The models in Tables 6.2a, 6.2b, and 6.2c indicate complexity in how cultural values configurations indicate need motivations, but the configurations are interpretable and provide support for the general connectedness of cultural values and general need motivations.

The findings that all but one of the models of cultural value configurations in the three parts of Tables 6.2a, 6.2b, and 6.2c include all seven value dimensions and that the models are quite useful in indicating high scores for nAch, nAff, nPow, ~nAch,

Table 6.2b For P1: Cultural value configurations indicating high nAff and the negation of nAff

(A) **Cultural values configuration model indicating high nAff**			
Model: aff_cc = f(egalitar_c, intel_auto_c, aff_auto_c, mast_c, hier_c, embed_c, harm_c)			
INTERMEDIATE SOLUTION: frequency cutoff: 1.00; consistency cutoff: 0.86.			
Model	**Raw coverage**	**Unique coverage**	**Consistency**
1. egalitar_c•intel_auto_c•~aff_auto_c•~mast_c•~hier_c•harm_c	0.16	0.16	0.85
Solution coverage: 0.16; solution consistency: 0.85			
(B) **Cultural values configuration model indicating negation of nAff**			
Model: not_aff_cc = f(egalitar_c, intel_auto_c, aff_auto_c, mast_c, hier_c, embed_c, harm_c)			
INTERMEDIATE SOLUTION: frequency cutoff: 1.00; consistency cutoff: 0.85			
Models	**Raw coverage**	**Unique coverage**	**Consistency**
1. ~egalitar_c•~intel_auto_c•aff_auto_c•mast_c•hier_c•embed_c•~harm_c	0.09	0.06	0.85
2. egalitar_c•intel_auto_c•~aff_auto_c•~mast_c•~hier_c•embed_c•harm_c	0.08	0.03	0.84
3. egalitar_c•intel_auto_c•~aff_auto_c•mast_c•hier_c•~embed_c•harm_c	0.11	0.05	1.00
Solution coverage: 0.20; solution consistency: 0.87			

Table 6.2c For P1: Cultural value configuration indicating high nPow and the negation of nPow

(A) **Cultural values configurations indicating high nPow**			
Model: nPow = f(egalitar_c, intel_auto_c, aff_auto_c, mast_c, hier_c, embed_c, harm_c)			
INTERMEDIATE SOLUTION: frequency cutoff: 1.00; consistency cutoff: 0.91			
Models	**Raw coverage**	**Unique coverage**	**Consistency**
1. ~egalitar_c•~intel_auto_c•aff_auto_c•mast_c•hier_c•embed_c•~harm_c	0.11	0.09	0.91
2. egalitar_c•intel_auto_c•~aff_auto_c•~mast_c•~hier_c•embed_c•harm_c	0.09	0.08	0.97
Solution coverage: 0.18; solution consistency: 0.95.			
(B) **Cultural values configurations indicating low nPow**			
Model: ~nPow = f(egalitar_c, intel_auto_c, aff_auto_c, mast_c, hier_c, embed_c, harm_c)			
INTERMEDIATE SOLUTION: frequency cutoff: 1.00; consistency cutoff: 0.88			
Models	**Raw coverage**	**Unique coverage**	**Consistency**
1. egalitar_c•intel_auto_c•aff_auto_c•~mast_c•~hier_c•~embed_c•~harm_c	0.18	0.02	0.94
2. egalitar_c•intel_auto_c•aff_auto_c•mast_c•~hier_c•~embed_c•harm_c	0.36	0.19	0.94
3. egalitar_c•intel_auto_c•~aff_auto_c•mast_c•hier_c•~embed_c•harm_c	0.10	0.06	0.88
Solution coverage: 0.44; solution consistency: 0.91.			

~nAff, and ~nPow provides strong support for applying Schwartz's theory as foundational for the study of need motivations. Also, the additional findings below extend this conclusion to the study of entrepreneur behavior, quality of life, and corruption.

The findings also support the usefulness of complexity tenets in examining how the seven value dimensions impact need motivations. Using symmetric tests to examine for simple directional relationships for each value for an outcome condition is insufficient when the directionality for a cultural value to predict an outcome is sometimes positive and sometimes negative—which depends on what additional conditions are present in the configurations. For example, different models in Table 6.2a include egalitarianism and the negation of egalitarianism to predict high scores in nAch. Comparing the models across the three parts of Tables 6.2a, 6.2b, and 6.2c results in similar observations for the other six cultural values—though egalitarianism and intellectual autonomy are high consistently, and embeddedness is low consistently in the models predicting the negation of nPow (Table 6.2c). The findings in Tables 6.2a, 6.2b, and 6.2c not only provide insights into the relevancy of formulating/testing Schwartz theory using complexity theory and asymmetric analysis, but the findings support the suggestion that directional relationship testing using symmetric analysis (e.g., MRA) is inadequate.

Consequently, Schwartz's analysis of using difference scores (e.g., autonomy minus embeddedness, egalitarianism minus hierarchy, and harmony minus mastery) and computing the correlation of differences with dependent measures—such as metrics for GDP, corruption levels, and democratization—is a bad science practice. Even if the correlations are statistically significant, the presence of high (low) autonomy can occur with low (high) embeddedness for high (low) GDP nations. The use of correlation analysis and MRA is shallow, and the findings obfuscate when high (low) cultural value scores are relevant for indicating a case high (low) for a simple, as well as complex, outcome such as high (low) GDP or high (low) corruption—or the four complex outcomes of high/low GDP by high/low corruption.

7.6 The Findings Support P2: Need Motivation Configurations Indicate Entrepreneur Behavior

The consistency indexes and XY plots for the simple conditions of each need motivation do *not* support the assertion that high nAch is a necessary condition for indicating high GEM, GEDI, and GII. The nine consistency indexes for the three need motivations by the three EB activities (GEM, GEDI, and GII) had consistency indexes within 0.40 and 0.60. The consistency indexes for the three need motivations indicating the negations of GEM, GEDI, and GII are within 0.40 and 0.75. Conclusion: the simple conditions of each need motivation do not provide high consistency indexes for indicating EB nor negation of EB.

A few complex configurations of need motivations do indicate high EB as well as the negation of EB. Table 6.3 and Fig. 6.5 include findings for need motivation configurations indicating high (low) GEM activity. One need motivation configuration (i.e., ~nAch•~nPow) indicates high GEM activity. This model is suggestive that high or low nAff is irrelevant for accurate prediction of high GEM—the negation of Ach with the negation of nPow in nations is sufficient for indicating high GEM no

Table 6.3 Need motivation models indicating nations with high (low) Global Entrepreneur Monitor (GEM) scores

(A) **High GEM Model: gem_c = f(pow_cc, aff_cc, ach_cc)**			
INTERMEDIATE SOLUTION: frequency cutoff: 1.00; consistency cutoff: 0.87.			
Model	**Raw coverage**	**Unique coverage**	**Consistency**
1. ~pow_cc•~ach_cc	0.50	0.50	0.85
Solution coverage: 0.50; solution consistency: 0.85			
(B) **Low GEM Model: not_gem_c = f(pow_cc, aff_cc, ach_cc)**			
INTERMEDIATE SOLUTION: frequency cutoff: 1.00; consistency cutoff: 0.81			
Models	**Raw coverage**	**Unique coverage**	**Consistency**
1. ~pow_cc•ach_cc	0.48	0.39	0.83
2. pow_cc•~aff_cc•~ach_cc	0.22	0.12	0.81
Solution coverage: 0.60; solution consistency: 0.81.			

Note: The findings do not include an XY plot for either low GEM model

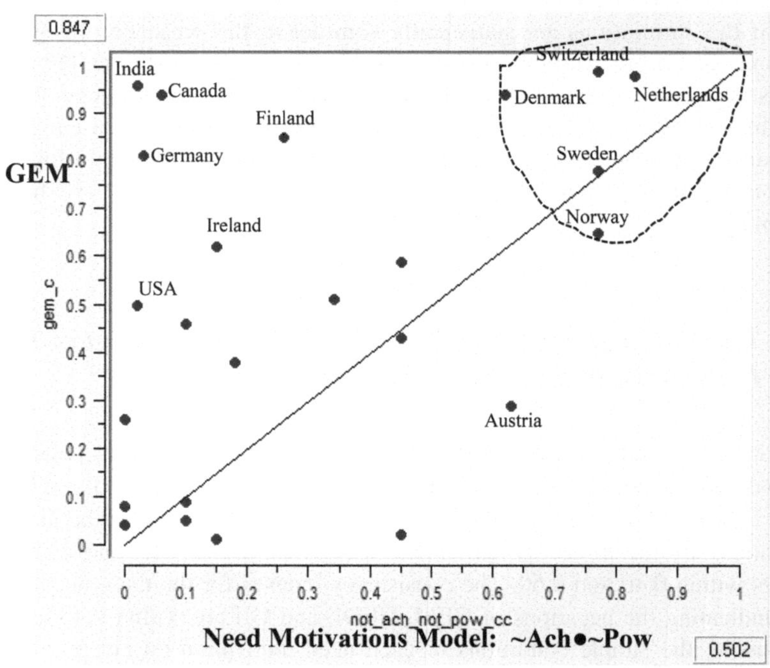

Fig. 6.5 Need motivation model indicating high GEM

matter the nations' scores on nAff. Table 6.3 and Fig. 6.5 include this model's high consistency (0.85). Figure 6.5 indicates that this model applies to northern European nations. The model indicating nations high in ~Ach•~nPow does not represent additional countries having high scores in GEM (e.g., Finland, Canada, Germany, and India). Thus, more than one need motivation model is necessary for indicating nations high in GEM.

Table 6.4a includes two complex configurations of need motivations indicating nations having high scores in GEDI with modestly high consistencies. Figure 6.6 identifies Switzerland and Sweden as two nations with high scores in model 1 in Table 6.4a (the negation of nAff AND the negation of nAch)—both these nations have high GEDI. Table 6.4b includes two complex configurations of need motivations indicating nations having high scores in the negation of GEDI with modestly high consistencies. Figure 6.7 identifies Greece and Portugal as the two nations with high scores in the complex statement ~nPow•nAff•nAch, model 2 in Table 6.4b—both these nations have high scores in the negation of GEDI. Such findings support the theoretical stance that the impacts of need motivations are complex but tractable. A few need motivation configurations support high GEDI, and a few need motivation configurations support the negation of GEDI.

This paper includes no additional tables for complex need motivation models indicating high GII or the negation of GII. Figure 6.4c shows the XY plot for a complex need motivation model (nAch•~nAff•nPow) indicating high GII activity. The following model is the only complex configuration of need motivations indicating high scores consistently (consistency = 0.81) for ~GII: ~nPow•nAch•nPow; this

Table 6.4 Need motivation models indicating nations with high (low) Global Entrepreneur Development Index (GEDI) scores

Table 4
Need motivation models indicating nations with high (low) Global Entrepreneur Development Index (GEDI) scores

A. High GEDI Model: gedi_c = f(pow_cc, aff_cc, ach_cc)
INTERMEDIATE SOLUTION: frequency cutoff: 1.00; consistency cutoff: 0.83.

Models	raw coverage	unique coverage	consistency
1. ~aff_cc•~ach_cc	0.37	0.26	0.83
2. pow_cc•aff_cc•ach_cc	0.26	0.15	0.87

Solution coverage: 0.52; solution consistency: 0.83.

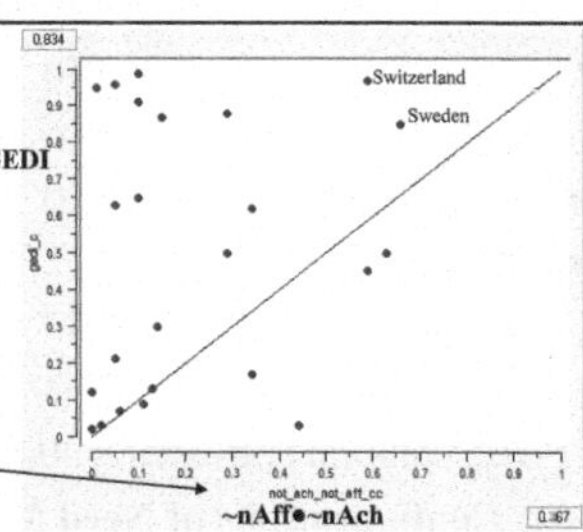

Figure 6.6
Need motivations model Indicating nations high in GEDI activity

B. Model: not_gedi_c = f(pow_cc, aff_cc, ach_cc)

INTERMEDIATE SOLUTION: frequency cutoff: 1.00; consistency cutoff: 0.86.

Models	raw coverage	unique coverage	consistency
1. pow•~aff	0.44	0.33	0.81
2. ~pow•aff_cc•ach	0.23	0.13	0.86

Solution coverage: 0.56; solution consistency: 0.82.

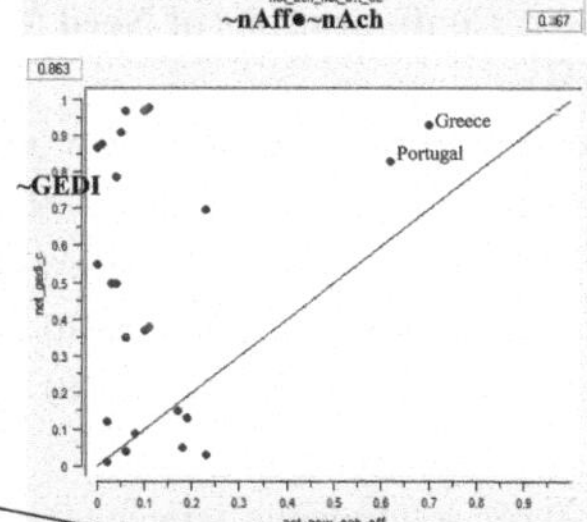

Figure 6.7
Need motivations model indicating nations high in ~GEDI activity

model indicates Greece and Portugal as the two nations high in the model, and both nations are high in the ~GII. Both of these nations have low levels EB (i.e., ~GEDI and ~GII). The GEM activities for these two nations are *not* high as well.

7.7 Do Complex Configurations of Need Motivations and GEM and GEDI Indicate High GII?

This study includes testing the suspicion that complex antecedent conditions that include need motivations and GEM and GEDI are accurate in predicting GEM and GEDI. Given the great importance in the relevant literature on innovation as a source of firm and national economic growth and success, achieving high GII activity may be a critical ingredient in recipes for nations high in QOL. (The findings in the present study indicate that GII alone has a modestly high consistency (0.81) in indicating nations with high QOL.)

The findings in Table 6.5 provide evidence that strongly support this hunch. High GEM or high GEDI or high scores in both in combination of specific need motivation configurations indicate nations having high GII scores consistently for six models in Table 6.5a. The negation of GEDI or the negation of GII, or both,

Table 6.5 Complex antecedent configurations of core motives and GEM and GEDI indicating High GII

(A) **MODEL: High Global Innovation Index = f (Need for Power, Need for Affiliation, Need for Achievement, Global Entrepreneur Index, Global Entrepreneur Monitor)**			
Model: gii_c = f(pow_cc, aff_cc, ach_cc, gedi_c, gem_c)			
Frequency cutoff: 1.00; consistency cutoff: 0.848			
Models	**Raw coverage**	**Unique coverage**	**Consistency**
1. aff_cc•~ach_cc•gem_c	0.46	0.08	0.95
2. ~pow_cc•~aff_cc•ach_cc•gedi_c	0.26	0.05	0.87
3. pow_cc•aff_cc•~ach_cc•gedi_c	0.30	0.04	1.00
4. pow_cc•aff_cc•gedi_c•gem_c	0.32	0.05	0.91
5. ~pow_cc•~ach_cc•gedi_c•gem_c	0.40	0.02	0.99
6. ~pow_cc•~aff_cc•gedi_c•gem_c	0.33	0.00	0.98
Solution coverage: 0.76; solution consistency: 0.89			
Complex Antecedent Configurations of Need Motivations and GEM and GEDI Indicating the Negation of GII			
(B) **Model: not_gii_c = f(gedi_c, gem_c, pow_cc, aff_cc, ach_cc)**			
Frequency cutoff: 1.00; consistency cutoff: 0.849.			
Models	**Raw coverage**	**Unique coverage**	**Consistency**
1. ~gedi_c•~gem_c•pow_cc•~ach_cc	0.30	0.19	0.90
2. ~gedi_c•~gem_c•~aff_cc•ach_cc	0.40	0.24	0.91
3. ~gedi_c•~pow_cc•aff_cc•ach_cc	0.23	0.08	0.91
Solution coverage: 0.68; solution consistency: 0.92.			

appears in models accurate in identifying nations low in GII in Table 6.5b. These findings and additional findings support the view that GEM and GEDI are insufficient by themselves in generating high QOL but both contribute along with specific need motivation ingredients to high GII which is sufficient for indicating high QOL. Also, the negation of GEM and the negation of GEDI in configurations with specific need motivations consistently indicate nations with low QOL. These findings support national policies nurturing GEM and GEDI activities and not just GII activities. However, national policy-makers and administrators should not rely generating high GEM, the apparent strategy implemented by India (cf. Singh and Seshadri 2012), since GEM alone is insufficient for indicating high QOL.

7.8 P3 and P4: Schwartz Cultural Value Configurations, Religiosity, and Need Motivations

The findings support P3 and P4. Including religiosity in configurations with Schwartz's cultural values serves to increase the consistencies of the models for indicating nAch, ~nAch, nAff, ~nAff, nPow, and ~nPow. While not sufficient by itself in identifying nations high (low) in each of the three need motivations, the inclusion of religiosity is an integral ingredient in all but one of the models indicating the six outcome conditions. Details for religiosity as an added ingredient with cultural values configurations for indicating each need motivation appear in Appendices D–E.

The Appendices D–F demonstrate that including religiosity in the cultural value configurations increases coverages of the models indicating high and low need motivations appearing in Tables 6.2a, 6.2b, and 6.2c. For indicating nAch, solution coverage jumps from 0.25 to 0.53 with the inclusion of religiosity in the cultural value configurations. For indicating the negation of nAch, solution coverage jumps from 0.17 to 0.41 with the inclusion of religiosity in the cultural value configurations. Similar jumps occur with the inclusion of religiosity in the cultural value configurations for nAff, ~nAff, and nPow. No jump occurs for ~nPow by including religiosity in the cultural value configurations. The cultural value configurations in Tables 6.2a, 6.2b, and 6.2c are sufficient for indicating the singular need motivations, but the addition of religiosity substantially improves the coverage of these models.

7.9 P5 Findings: Need Motivation Configurations Indicating High/Low QOL

The following need motivation configuration indicates high QOL with a 0.85 consistency and a 0.49 coverage: ~nAch•~nPow. The following need motivation configuration indicates the negation of QOL with a 0.88 consistency and a 0.25

coverage: nAch•nAff•~nPow. Testing the "Danish need motivation configuration" (i.e., ~nAch•nAff•~nPow) provides a 0.86 consistency and a 0.40 coverage for QOL. Testing the American need motivation configuration (i.e., nAch•nAff•nPow) provides a 0.78 consistency and 0.22 coverage—findings below the threshold of useful consistency. Figures 6.8 and 6.9 include additional details of findings for the Danish and American need motivation models, respectively.

The finding that ~nPow is an ingredient in both principal models for QOL and ~QOL is counterintuitive though illustrating a principal complexity theory tenet. The negation of the nPow in the Danish model may illustrate nations attempting to nurture members of their societies (i.e., nAff) without attempting to control their beliefs. Greece is representative of the finding that high ~nPow is an ingredient in the need motivation configuration indicating the negation of QOL. The main observation relevant to the fifth proposition (P5) is that need motivation configurations offer insights into QOL but are not particularly robust in the knowledge that they provide. The associations of need motivation configurations with EB are more informative than the associations of need motivation configurations with QOL. This conclusion fits the direct links of arrows 2 and 8 in Fig. 6.1. For arrow 8, the findings below for proposition 8 provide substantial support that high EB indicates high QOL accurately.

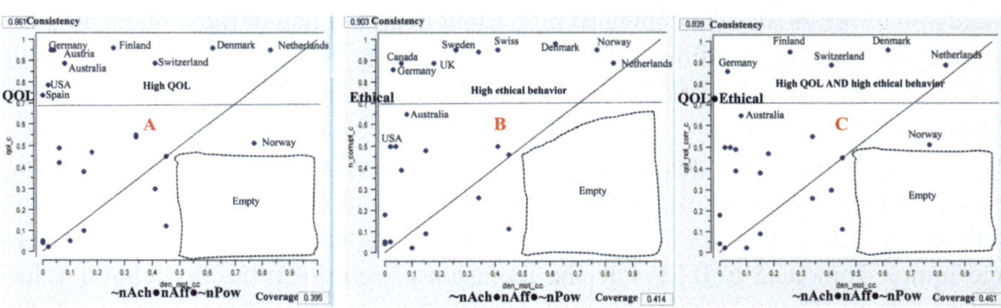

Fig. 6.8 Danish motives model (~nAch•nAff•~nPow) impacts on (**a**) national quality of life (QOL), (**b**) ~corruption, and (**c**) QOL•corruption. *Findings*: The Danish motives model has high consistencies for indicating high quality of life, high ethical behavior, and the high QOL•Ethical configuration

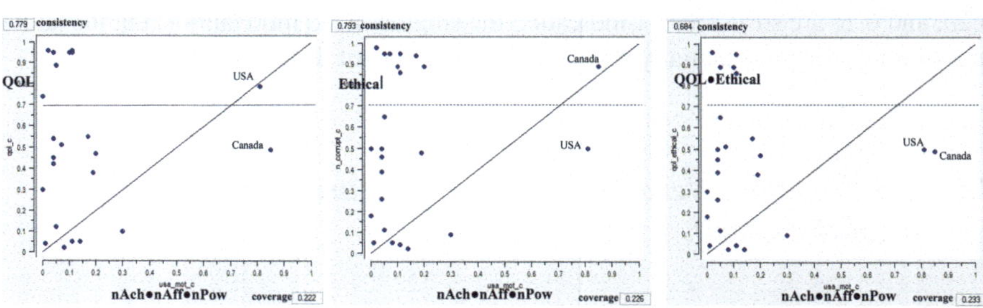

Fig. 6.9 USA's motives model (nAch•nAff•nPow) impacts on (**a**) national quality of life (QOL), (**b**) ~corruption, and (**c**) QOL•corruption. *Findings*: USA's motives model alone does not indicate high consistencies for QOL, ethical behavior, or QOL•ethical behavior

Two need motivation configurations indicate nations with high ethical behavior. Two need motivation models indicate nations with high corruption. Appendix C includes the details for how these additional findings support P5.

7.10 *P6 Findings: Cultural Value Configurations Indicate High (Low) Entrepreneur Behavior*

This section presents findings for P6: cultural values configurations (including religiosity) indicate nations with high (low) scores in entrepreneur behavior of GEM, GEDI, and GII. The analysis includes separate examinations for each of these three EBs. Table 6.6 presents the findings for cultural value configurations indicating nations (a) high and (b) low in GEM. Note in part A of Table 6.6 that affective

Table 6.6 Nations' cultural value configurations indicating high/low Global Entrepreneurship Monitor (GEM) scores

(A) **Cultural Value Configurations Indicating High GEM: Model: gem_c = f(relig_c, egalitar_c, intel_auto_c, aff_auto_c, mast_c, hier_c, embed_c, harm_c)**			
INTERMEDIATE SOLUTION: frequency cutoff: 1.00; consistency cutoff: 0.84.			
Models	**Raw coverage**	**Unique coverage**	**Consistency**
1. relig_c•egalitar_c•~intel_auto_c•aff_auto_c•mast_c•hier_c•~embed_c•~harm_c	0.11	0.03	0.84
2. relig_c•~egalitar_c•~intel_auto_c•aff_auto_c•mast_c•hier_c•embed_c•~harm_c	0.10	0.04	0.94
3. ~relig_c•egalitar_c•intel_auto_c•aff_auto_c•mast_c•~hier_c•~embed_c•harm_c	0.21	0.17	0.86
Solution coverage: 0.31; solution consistency: 0.87.			
(B) **Cultural Value Configurations Indicating Negation of GEM: Model: not_gem_c = f(relig_c, egalitar_c, intel_auto_c, aff_auto_c, mast_c, hier_c, embed_c, harm_c)**			
INTERMEDIATE SOLUTION: frequency cutoff: 1.00; consistency cutoff: 0.84.			
Models	**Raw coverage**	**Unique coverage**	**Consistency**
1. relig_c•egalitar_c•intel_auto_c•~mast_c•~hier_c•~embed_c•harm_c	0.35	0.18	0.87
2. relig_c•egalitar_c•intel_auto_c•~aff_auto_c•~mast_c•~hier_c•harm_c	0.18	0.02	1.00
3. relig_c•egalitar_c•~intel_auto_c•aff_auto_c•mast_c•hier_c•~embed_c•~harm_c	0.12	0.05	0.87
4. relig_c•egalitar_c•intel_auto_c•~aff_auto_c•mast_c•hier_c•~embed_c•harm_c	0.12	0.04	0.98
Solution coverage: 0.48 solution consistency: 0.88.			

Notes: The models do support the proposition that different cultural value configurations indicate GEM and ~GEM. The models have high consistency (accuracy)

Religiosity is an ingredient in all models for GEM and ~GEM. Both harmony and ~harmony are ingredients in some of the models for GEM and ~GEM autonomy and mastery are both ingredients in all three models. No other cultural value is consistently positive or negative among these three models for high GEM. Affective autonomy with mastery serves as a core building block for constructing cultural value configurations for high GEM—necessary but insufficient ingredients in these recipes. However, both affective autonomy and mastery together are ingredients in model 3 for the negation of GEM in the part B of Table 6.6. Such findings indicate the folly on relying on configurations of two or three ingredients in complex antecedent conditions in predicting outcomes.

Table 6.7 presents the findings for cultural values configurations indicating high (low) GEDI. The findings indicate that cultural values configurations are indeed useful for indicating high GEDI as well as the negation of GEDI. Similar to the findings for cultural value configurations and GEM, affective autonomy and mastery together

Table 6.7 Nations' cultural value configurations indicating high/low global Entrepreneurship Development Index (GEDI) scores

(A) **Cultural Value Configurations Indicating High GEDI: Model: gedi_c = f(relig_c, egalitar_c, intel_auto_c, aff_auto_c, mast_c, hier_c, embed_c, harm_c)**			
INTERMEDIATE SOLUTION: frequency cutoff: 1.00; consistency cutoff: 0.88.			
Models	**Raw coverage**	**Unique coverage**	**Consistency**
1. egalitar_c•intel_auto_c•aff_auto_c•mast_c•~hier_c•~embed_c•~harm_c	0.35	0.08	0.93
2. ~relig_c•egalitar_c•intel_auto_c•aff_auto_c•mast_c•~embed_c•~harm_c	0.30	0.04	0.98
3. ~relig_c•egalitar_c•intel_auto_c•aff_auto_c•mast_c•~hier_c•~embed_c	0.35	0.09	0.99
4. relig_c•egalitar_c•~intel_auto_c•aff_auto_c•mast_c•hier_c•~embed_c•~harm_c	0.12	0.03	0.88
Solution coverage: 0.51; solution consistency: 0.94.			
(B) **Cultural Value Configurations Indicating Low GEDI: Model: not_gedi_c = f(relig_c, egalitar_c, intel_auto_c, aff_auto_c, mast_c, hier_c, embed_c, harm_c)**			
INTERMEDIATE SOLUTION: frequency cutoff: 1.00; consistency cutoff: 0.82.			
Models	**Raw coverage**	**Unique coverage**	**Consistency**
1. relig_c•egalitar_c•intel_auto_c•~aff_auto_c•~mast_c•~hier_c•harm_c	0.17	0.08	1.00
2. relig_c•egalitar_c•intel_auto_c•aff_auto_c•~hier_c•~embed_c•harm_c	0.38	0.28	0.86
3. ~relig_c•~egalitar_c•intel_auto_c•aff_auto_c•mast_c•hier_c•~embed_c•harm_c	0.11	0.07	0.82
4. relig_c•egalitar_c•intel_auto_c•~aff_auto_c•mast_c•hier_c•~embed_c•harm_c	0.11	0.05	1.00
Solution coverage: 0.59; solution consistency: 0.89.			

Notes: The models do support the proposition that different cultural value configurations indicate GEDI and ~GEDI. The models have high consistency (accuracy)

Religiosity and ~religiosity are ingredients for one or more models for GEDI and ~GEDI. The negation of harmony is an ingredient in 3 of 4 models for GEDI

Harmony is an ingredient in 4 of 4 models for ~GEDI. These findings support the perspective that entrepreneurship activity is disruptive appear in four of four models in Table 6.7, part A, and in model 3 among the four models in part B. Since GII activity is disruptive by its focus on change, harmony would be expected to be negative as an ingredient in models indicating high GEDI; the findings support this sub-proposition—the negation of harmony appears in three of the four models, and harmony/~harmony is absent from model 3 in Table 6.7, part A. Harmony appears in all four models in the negation of GEM, Table 6.7, part B. Given harmony relates to static behavior, this finding provides nomological validity to the perspective that cultural value configurations indicate GEDI and the negation of GEDI.

The findings demonstrate that cultural value configurations indicate nations having high (low) scores for GII. Table 6.8 presents details. Affective autonomy with

Table 6.8 Nations' cultural value configurations indicating high/low Global Innovation Index (GII) scores

(A) **Cultural Value Configurations Indicating High GII: Model: gii_c = f(relig_c, egalitar_c, intel_auto_c, aff_auto_c, mast_c, hier_c, embed_c, harm_c)**			
INTERMEDIATE SOLUTION: frequency cutoff: 1.00; consistency cutoff: 0.86.			
Models	**Raw coverage**	**Unique coverage**	**Consistency**
1. egalitar_c•intel_auto_c•aff_auto_c•mast_c•~hier_c•~embed_c•~harm_c	0.29	0.05	0.83
2. ~relig_c•egalitar_c•intel_auto_c•aff_auto_c•mast_c•~embed_c•~harm_c	0.26	0.03	0.95
3. ~relig_c•egalitar_c•intel_auto_c•aff_auto_c•~hier_c•~embed_c•harm_c	0.55	0.40	0.90
4. relig_c•egalitar_c•~intel_auto_c•aff_auto_c•mast_c•hier_c•~embed_c•~harm_c	0.11	0.02	0.86
5. ~relig_c•~egalitar_c•intel_auto_c•aff_auto_c•mast_c•hier_c•~embed_c•harm_c	0.11	0.05	0.88
Solution coverage: 0.80; solution consistency: 0.85.			
(B) **Cultural Value Configurations Indicating the Negation of GII: Model: not_gii_c = f(relig_c, egalitar_c, intel_auto_c, aff_auto_c, mast_c, hier_c, embed_c, harm_c)**			
INTERMEDIATE SOLUTION: frequency cutoff: 1.00; consistency cutoff: 0.84.			
Models	**Raw coverage**	**Unique coverage**	**Consistency**
1. relig_c•egalitar_c•intel_auto_c•~aff_auto_c•~mast_c•~hier_c•harm_c	0.19	0.12	1.00
2. relig_c•~egalitar_c•~intel_auto_c•aff_auto_c•mast_c•hier_c•embed_c•~harm_c	0.10	0.05	0.84
3. lrelig_c•egalitar_c•intel_auto_c•aff_auto_c•mast_c•~hier_c•~embed_c•harm_c	0.28	0.20	0.88
4. relig_c•egalitar_c•intel_auto_c•~aff_auto_c•mast_c•hier_c•~embed_c•harm_c	0.12	0.04	0.99
Solution coverage: 0.52; solution consistency: 0.91.			

Notes: The models do support the proposition that different cultural value configurations indicate GII and ~GEII. The models have high consistency (accuracy)
The negation of religiosity and religiosity are ingredients in 4 of the 5 models for GII. Religiosity is an ingredient in 4 of 4 models for the negation of GII
The negation of harmony is an ingredient in 4 of 5 models indicating GII and harmony is an ingredient in 3 of 4 models for the negation of GII

mastery is an ingredient in five of the five models in Table 6.8, part A. Affective autonomy is an ingredient in all five models, and mastery/~mastery is absent in one of the five models. These findings support and help to generalize the perspective that affective autonomy and mastery are both ingredients in most models indicating high GEM, GEDI, and GII—though by itself, the combination of affective autonomy and mastery is not necessary nor sufficient for any of the three EB measures as outcomes. Note in Table 6.8, part B, that affective autonomy and mastery appear together in models 2 and 3 for indicating the negation of GII.

The negation of harmony is an ingredient in three of the five models in Table 6.8, part A. Harmony is in the remaining two models in Table 6.8, part B. The opposite pattern appears in Table 6.8, part B, among the models indicating the negation of GII: three of four models include harmony, and one includes the negation of harmony. Religiosity is an ingredient in all the models indicating the negation of GII with high accuracy (Table 6.8, part B).

7.11 The Findings Support P7a: Religiosity Embeds in Cultural Values Configurations Indicates High (Low) QOL

The findings for P7 for cultural values plus religiosity indicating nAch appear in Table 6.9 for both high QOL (Table 6.9a) and low QOL nations (Table 6.9b). The findings include high or low religiosity in all four models indicating high QOL. The finding that an ingredient (antecedent condition) can have both a positive and a negative impact on an outcome condition is one of the central tenets of complexity theory. The directionality of impact of an antecedent condition depends on the additional ingredients present and their directionalities in the complex configurations. Note in Table 6.9b that high religiosity is an ingredient in all four models associating with ~QOL. Altogether, these findings indicate a complex impact of religiosity on nations' QOL. Viewing high (low) religiosity as a positive (or negative) force on a nation's QOL is a perspective too shallow to be accurate or useful.

The presence of either six or seven cultural values in all the models in Table 6.9a, b is a remarkable finding supporting the usefulness of Schwartz core conceptualization of cultural values configurations affecting human behavior. For most (not all) of the two sets of models in Table 6.9 (a and b models), Schwartz's (2018) perspective receives support: most of the models include the Boolean algebra representations analogous to Schwartz matrix algebra calculations of autonomy minus embeddedness, egalitarianism minus hierarchy, and/or harmony minus mastery. However, the use of set and complexity theory with Boolean algebra avoids the problems (Peter et al. 1993; Teas 1993) inherent with computing difference scores and building theory based on difference scores. The findings in the present study serve to deepen as well as support the relevance of Schwartz seven cultural values—in configurations by themselves or with religiosity—in indicating high as well as low QOL.

Table 6.9 Cultural values and religiosity indicating high and low QOL

(A) **Cultural values configurations plus religiosity indicating high QOL**			
Model: qol_c = f(relig_c, egalitar_c, intel_auto_c, aff_auto_c, mast_c, hier_c, embed_c, harm_c)			
INTERMEDIATE SOLUTION: frequency cutoff: 1.00; consistency cutoff: 0.85			
Models	**Raw coverage**	**Unique coverage**	**Consistency**
1. ~relig_c•egalitar_c•intel_auto_c•aff_auto_c•mast_c•~embed_c•~harm_c	0.26	0.02	0.91
2. ~relig_c•egalitar_c•intel_auto_c•aff_auto_c•mast_c•~hier_c•~embed_c	0.33	0.04	0.99
3. relig_c•egalitar_c•~intel_auto_c•aff_auto_c•mast_c•hier_c•~embed_c•~harm_c	0.11	0.04	0.85
4. relig_c•egalitar_c•intel_auto_c•aff_auto_c•~mast_c•~hier_c•~embed_c•harm_c	0.26	0.14	0.87
Solution coverage: 0.52; solution consistency: 0.88.			
(B) **Cultural values configurations plus religiosity indicating low QOL**			
Model: not_qol_c = f(relig_c, egalitar_c, intel_auto_c, aff_auto_c, mast_c, hier_c, embed_c, harm_c)			
INTERMEDIATE SOLUTION: frequency cutoff: 1.00; consistency cutoff: 0.82			
Models	**Raw coverage**	**Unique coverage**	**Consistency**
1. relig_c•egalitar_c•intel_auto_c•~aff_auto_c•~mast_c•~hier_c•harm_c	0.18	0.12	1.00
2. relig_c•egalitar_c•intel_auto_c•aff_auto_c•~mast_c•~hier_c•~embed_c•~harm_c	0.12	0.07	0.82
3. relig_c•~egalitar_c•~intel_auto_c•aff_auto_c•mast_c•hier_c•embed_c•~harm_c	0.11	0.06	0.92
4. relig_c•egalitar_c•intel_auto_c•~aff_auto_c•mast_c•hier_c•~embed_c•harm_c	0.12	0.05	1.00
Solution coverage: 0.39; solution consistency: 0.93.			

7.12 *The Findings Support P7B: Religiosity Embeds in Cultural Values Configurations Indicates High Ethical Behavior or High Corruption*

Details of the findings from testing P7 appear in Table 6.10 and Fig. 6.10. Part A in Table 6.10 includes three models indicating nations high in ethical behavior. Two of these three models include the negation of religiosity, and one includes religiosity. With consistencies equal or above 0.85, all three models are useful for indicating high ethical behavior. Coverage of nations high in ethical behavior is quite high (coverage = 0.74).

Part B in Table 6.10 includes six models indicating nations high in corruption. All six models include religiosity with cultural values configurations. Among the seven or eight conditions in the six models, only high religiosity appears in all six

Table 6.10 Cultural values and religiosity indicating high ethical behavior versus high corruption

(A) **Cultural values and religiosity indicating high ethical behavior: Model: n_corrupt_c = f(relig_c, egalitar_c, intel_auto_c, aff_auto_c, mast_c, hier_c, embed_c, harm_c)**			
INTERMEDIATE SOLUTION: frequency cutoff: 1.00; consistency cutoff: 0.86.			
Models	**Raw coverage**	**Unique coverage**	**Consistency**
1. ~relig_c∗egalitar_c∗intel_auto_c∗aff_auto_c∗mast_c∗~embed_c∗~harm_c	0.28	0.10	0.99
2. ~relig_c∗egalitar_c∗intel_auto_c∗aff_auto_c∗~hier_c∗~embed_c∗harm_c	0.57	0.41	0.90
3. relig_c∗egalitar_c∗~intel_auto_c∗aff_auto_c∗mast_c∗hier_c∗~embed_c∗~harm_c	0.11	0.04	0.85
Solution coverage: 0.74; solution consistency: 0.91.			
(B) **Cultural values and religiosity indicating high corruption: Model: corrupt = f(relig_c, egalitar_c, intel_auto_c, aff_auto_c, mast_c, hier_c, embed_c, harm_c)**			
INTERMEDIATE SOLUTION: frequency cutoff: 1.00; consistency cutoff: 0.83.			
Models	**Raw coverage**	**Unique coverage**	**Consistency**
1. relig_c∗egalitar_c∗intel_auto_c∗~aff_auto_c∗~mast_c∗~hier_c∗harm_c	0.18	0.11	1.00
2. relig_c∗egalitar_c∗intel_auto_c∗aff_auto_c∗~mast_c∗~hier_c∗~embed_c∗~harm_c	0.12	0.03	0.83
3. relig_c∗egalitar_c∗~intel_auto_c∗aff_auto_c∗mast_c∗hier_c∗~embed_c∗~harm_c	0.14	0.03	0.95
4. relig_c∗~egalitar_c∗~intel_auto_c∗aff_auto_c∗mast_c∗hier_c∗embed_c∗~harm_c	0.12	0.04	1.00
5. relig_c∗egalitar_c∗intel_auto_c∗aff_auto_c∗mast_c∗~hier_c∗~embed_c∗harm_c	0.28	0.16	0.92
6. relig_c∗egalitar_c∗intel_auto_c∗~aff_auto_c∗mast_c∗hier_c∗~embed_c∗harm_c	0.12	0.04	1.00
Solution coverage: 0.58; solution consistency: 0.95.			

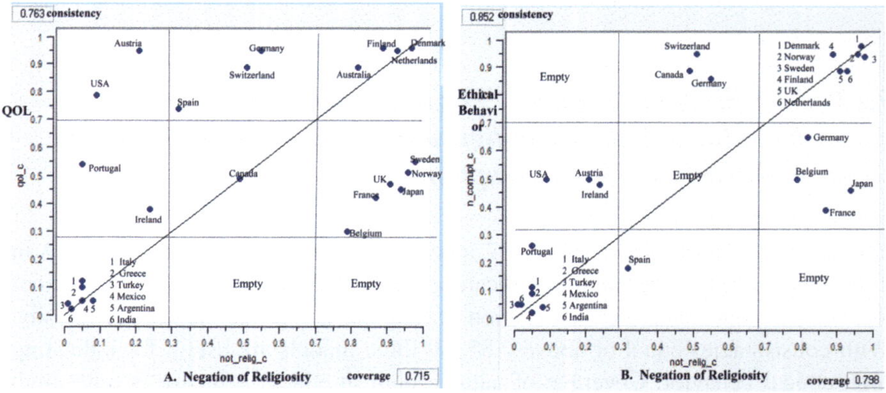

Fig. 6.10 Negation of religiosity indicating QOL and ethical behavior. Notes: (1) All nations low in QOL are high in religiosity (i.e., low in negation of religiosity). (2) Note that being high in the negation of religiosity does not indicate high QOL consistently: The USA and Austria are high in religiosity and high in QOL. (3) All nations low in religiosity have high or moderate levels of ethical behavior. (4) Nearly all nations high in religiosity have high corruption scores (i.e., low in the negation of corruption)

models. Most but not all nations high in religiosity have high corruption scores, but all nations high in corruption (except Spain) are nations high in religiosity.

Part A of Fig. 6.10 indicates that nations high in negation of religiosity do not have high QOL scores consistently (763). Some do but some do not. Austria and the USA achieve high QOL, but also these nations are low in the negation of religiosity (i.e., they have high religiosity). Part B of Fig. 6.10 indicates that nations high in the negation of religiosity do have high ethical behavior scores consistently (0.852). A high score in religiosity weakly indicates a nation high in corruption (consistency = 0.793)—this paper does not include the XY plot since the reverse of Fig. 6.10, part B, indicates that all nations low in the negation of religiosity are low in ethical behavior except for Spain.

7.13 *The Findings Support P8: GEM, GEDI, and GII Models Indicate High (Low) QOL*

P8 includes proposing the complex antecedent configurations of EB indicating high and low QOL:

$$\text{GEM} \bullet \text{GEDI} \bullet \text{GII} \leq \text{QOL} \tag{6.4}$$

$$\sim \text{GEM} \bullet \sim \text{GEDI} \bullet \sim \text{GII} \leq \sim \text{QOL} \tag{6.5}$$

The detailed findings from testing these two models appear in Figs. 6.10 and 6.11. The findings support both models.

Nations expressing high EB in all three scores for GEM, GEDI and GII have high scores consistently in QOL. Figure 6.11 shows that the consistency index is 0.912 for this model. No nation having high scores in model 4 has a low QOL. Six of the nations having low scores in EB have low QOL. None of the nations with low QOL have high EB. Three nations manage to have high QOL even though they have low EB scores: Australia, Austria, and Spain.

Figure 6.12 shows that five of nine nations with high scores in the negation of EB have high scores in the negation of QOL. None of the remaining four nations with high scores in EB have low scores in ~QOL. One nation manages to have a low score in the negation of EB and a high score in the negation of QOL: India. India achieves this distinction by having a very active national program in nurturing GEM. However, India has low scores in GEDI and GII. The pattern of findings for India supports the view that India's high GEM efforts are doomed to fail without engaging in policies to nurture GEDI and GII.

Table 6.11 provides additional details of findings for models attempting to demonstrate that EB activity combines with need motivation configurations to indicate high, as well as low, QOL. Such modeling helps to deepen understanding regarding when nAch and the negation of nAch are ingredients in configurations supporting high versus low QOL. Note in Table 6.11 that both nAch and ~nAch are ingredients in different models indicating both QOL and ~QOL. The same observation holds

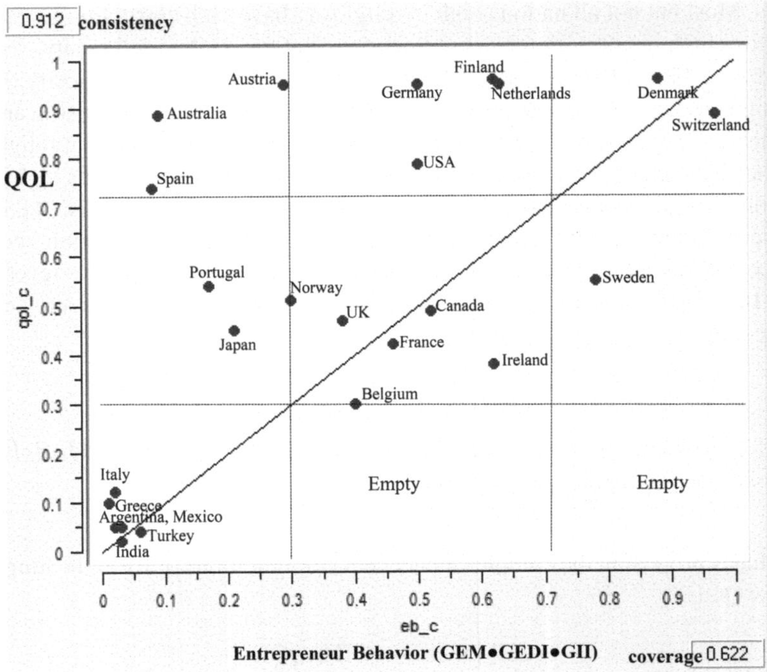

Fig. 6.11 Entrepreneur behavior (GEM•GEDI•GII) indicating high quality of life. *Findings*: Consistency of this model is quite high (0.912). Coverage is high (0.622). Nine nations have high scores in QOL. Two of three nations high in EB are high in QOL. No nation high in EB is low in QOL. Nations low in EB are low and high in QOL. Six nations low in EB are low in QOL. Three nations low in EB are high in QOL. The EB and QOL association is asymmetric. A nation high in EB is not necessary for achieving high QOL, but being high in EB is sufficient for achieving high or medium QOL. The odds are 2 to 1 that a nation will be low versus high in QOL if the nation is low in EB

for nAff and ~nAff, nPow and ~nPow, GEM and ~GEM, GEDI and ~GEDI, and GII and ~GII.

Except for model 7 in Table 6.11, part A, all the models in Table 6.11 have at least one negation condition. This observation holds from both the models indicating QOL and the negation of QOL. Given the associations contrary to ex ante expectations almost always occur, such deep examination of recipes is necessary to solve seemingly intractable antecedent-outcome findings—findings, for example, of nations low in nAch with high QOL.

Note in Table 6.11, part A, that six of the seven models for QOL include high GEDI; one model includes ~GEDI. In Table 6.11, part B, four of the six models include ~GEDI, and two include GEDI for indicating ~QOL. Most of the models with GEM and GII as ingredients for high QOL include high GEM and high GII, though the ~GEM or the ~GII does occur for one or two models in Table 6.10, part A. The negation of GII occurs in only one model in Table 6.11, part A, but occurs in four of the six models for ~QOL in Table 6.11, part B.

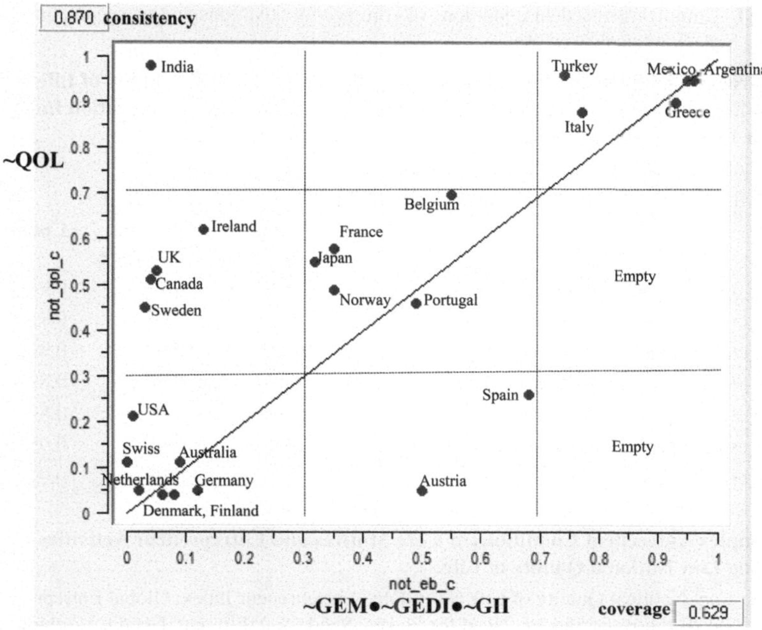

Fig. 6.12 Negation of entrepreneur behavior (~GEM•~GEDI•~GII) indicating low quality of life. *Findings*: Consistency of this model is quite high (0.870). Coverage is high (0.629). Six nations have high scores in ~QOL. All five nations with high scores in the negation of EB (~GEM•~GEDI•~GII) have high scores in ~QOL. This model is highly accurate (consistency = 0.87) in indicating nations high in the negation of QOL. These findings support the nurturing of EB by national laws and policies. India has a high score in GEM but low scores in GEDI and GII. The findings support the perspective that India policy leaders fail to recognize the high GEM alone is doomed to not nurture high QOL

7.14 The Findings Support P9: Nations High in Ethical Behavior Are High in QOL

Figure 6.13 is the XY plot for ethical behavior (i.e., the negation of corruption) and QOL. The 0.835 consistency index with 0.835 coverage index supports the conclusion that high ethical behavior indicates high QOL. No nation high in ethical behavior has low QOL. Five nations high in ethical behavior are high in QOL: Denmark, Finland, Germany, the Netherlands, and Switzerland. Spain manages to achieve high QOL with low ethical behavior. How Spain maintains this contrarian position is worthy of separate study and reporting.

In achieving moderate QOL, for four nations (Canada, Norway, Sweden, and UK) do not receive the full benefits of high ethical behavior. How these four nations might be able to boost their QOL performances while maintaining their high ethical behavior may associate with their moderate commitment to entrepreneur behavior.

Table 6.11 Complex antecedent conditions of core need motivations and entrepreneur activities indicating high (low) quality of life.

(A) **Entrepreneur Behavior Indicating High QOL: MODEL: High Quality of Life = f (Global Entrepreneur Index, Global Entrepreneur Monitor, Global Innovation Index, Need for Power, Need for Affiliation, Need for Achievement)**			
Model: qol_c = f(gedi_c, gem_c, gii_c, pow_cc, aff_cc, ach_cc)			
INTERMEDIATE SOLUTION: Frequency Cutoff: 1.00; Consistency Cutoff: 0.91			
Models	**Raw coverage**	**Unique coverage**	**Consistency**
1. ~gedi_c•gem_c•~gii_c•~pow_cc•aff_cc	0.19	0.05	0.94
2. gedi_c•~gem_c•~pow_cc•~aff_cc•ach_cc	0.24	0.07	0.93
3. gedi_c•gem_c•gii_c•~pow_cc•~ach_cc	0.38	0.07	0.94
4. gedi_c•gem_c•gii_c•~pow_cc•~aff_cc	0.30	0.00	0.89
5. gedi_c•gii_c•pow_cc•aff_cc•~ach_cc	0.28	0.01	0.89
6. gem_c•gii_c•pow_cc•aff_cc•~ach_cc	0.28	0.01	0.95
7. gedi_c•gem_c•gii_c•pow_cc•aff_cc	0.32	0.05	0.97
Solution coverage: 0.71; solution consistency: 0.88			
(B) **Complex Antecedent Conditions of Core Motives and Entrepreneur Activities Indicating Low National Quality of Life:**			
MODEL: Low National Quality of Life = f (Global Entrepreneur Index, Global Entrepreneur Monitor, Global Innovation Index, Need for Power, Need for Affiliation, Need for Achievement)			
Model: not_qol_c = f(gedi_c, gem_c, gii_c, pow_cc, aff_cc, ach_cc)			
INTERMEDIATE SOLUTION: Frequency Cutoff: 1.00; Consistency Cutoff: 0.84.			
Models	**Raw coverage**	**Unique coverage**	**Consistency**
1. ~gedi_c•~gem_c•~gii_c•pow_cc•~ach_cc	0.28	0.12	0.96
2. ~gedi_c•~gem_c•~gii_c•~pow_cc•ach_cc	0.32	0.14	0.90
3. ~gedi_c•gem_c•~gii_c•~pow_cc•aff_cc	0.21	0.02	0.94
4. ~gedi_c•gem_c•~gii_c•pow_cc•~aff_cc•ach_cc	0.16	0.08	0.94
5. gedi_c•~gem_c•gii_c•pow_cc•aff_cc•~ach_cc	0.21	0.06	0.89
6. gedi_c•gem_c•gii_c•~pow_cc•~aff_cc•ach_cc	0.19	0.06	0.83
Solution coverage: 0.75; solution consistency: 0.91			

Six nations are located in the quagmire of low ethical behavior and low QOL: Argentina, Greece, India, Italy, Mexico, and Turkey. National governments and groups protesting on the streets sometimes advocate the reduction/elimination of corruption as a necessary step for increasing QOL. For example, “A New Revolution in Mexico: Sick of corruption and of Trump, voters embrace the maverick leftist, Andrés Manuel López Obrador” (Anderson, 2018). However, though the asymmetric model has high consistency, the XY plot in Fig. 6.13 indicates that increasing QOL with low ethical behavior may be possible, Spain being an example. Possibly, a two- or multiple-front strategy is necessary to reduce corruption and increase QOL by additional steps. One conclusion that is likely useful is while high ethical behavior indicates nations high in QOL, the reduction in corruption is unlikely to be sufficient alone for achieving high QOL. For example, most of the nations that are high in QOL have moderate-to-high entrepreneur behavior.

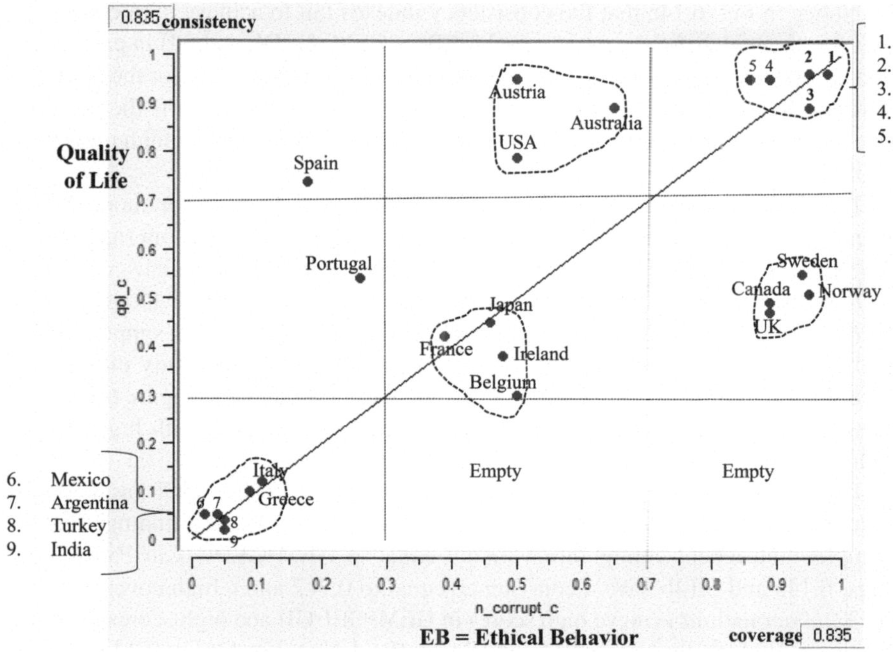

Fig. 6.13 Nation's ethical behavior indicating nation's quality of life (QOL). *Note*: The findings include modestly a high consistency (0.835) for the simple condition model (EB ≤ QOL). None of the nations low in EB are high in QOL. Five of the nine nations high in EB are high in QOL, and four achieve medium QOL levels; no nation high in EB is low in QOL

7.15 Entrepreneur Behavior Predicting the Complex Outcome: QOL•Ethical Behavior

Nearly all studies on nations' cultural values, need motivations, and entrepreneur behavior examine the impact of these conditions for simple outcomes, for example, quality of life separately from ethical behavior. However, examining the impact on simple and complex antecedent conditions on complex outcomes is possible and desirable. The present study includes examining simple and complex entrepreneur behaviors on the complex outcome: QOL AND ethical behavior. Figure 6.14a presents the findings.

The findings in Fig. 6.14a support the speculation that the weak commitment to EB contributes to the failure by Canada, Norway, Sweden, and the UK in achieving high QOL—even though these nations have high ethical behavior. These four nations have stuck-in-the-middle results for QOL AND ethical behavior in regard to each of the three EB antecedents— GEM, GEDI, and GII—as well as the complex antecedent condition, GEM•GEDI•GII. Fig. 6.14a presents these findings in sub-Figs. A, B, C, and D, respectively.

Notice in Fig. 6.14a that the consistency indexes fail to achieve high consistencies for the models for the three singular EBs—GEM, GEDI, and GII in parts A, B, and C. The consistency index indicates high accuracy (0.908) only for the complex antecedent condition equal to GEM•GEDI•GII. This finding supports the perspective that nations must support all three types of EB for EB to nurture high impact on QOL•Ethical.

India does achieve high GEM assessment and has done so for more than a decade. Notice in part A of Fig. 6.14a, India sits uniquely in the bottom right in this XY plot. India's weak performances in GEDI and GII associate with low QOL•Ethical. India's effort to spur the development of QOL via developing GEM is likely wasted motion and money because of the nation's failure to support GEDI and GII adequately. India appears to suffer from a double whammy effect: low levels of GEDI and GII along with low levels of ethical behavior. Figure 6.14a part D supports the perspective that high GEM•GEDI•GII coupled with high ethical indicates high QOL.

Figure 6.14b provides additional support for this conclusion. QOL, as a simple outcome condition, appears in the four panels of Fig. 6.14b. The findings for the complex antecedent configuration model equal to GEM•GEI•GII in Panel D in Figs. 6.14a and 6.14b have a consistency equal to 0.912 and a high coverage (i.e., 0.622). Four nations achieve high scores in GEM•GEI•GII and high scores in QOL: Denmark, Finland, the Netherlands, and Switzerland. No nation low in GEM•GEI•GII is high in QOL. The relationship between GEM•GEI•GII and QOL is asymmetric:

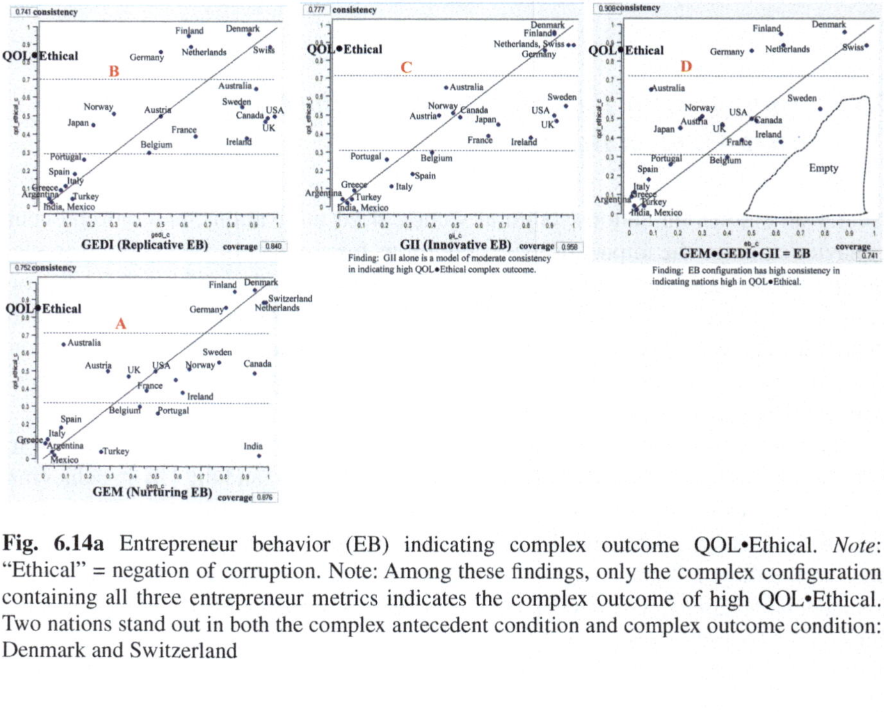

Fig. 6.14a Entrepreneur behavior (EB) indicating complex outcome QOL•Ethical. *Note*: "Ethical" = negation of corruption. Note: Among these findings, only the complex configuration containing all three entrepreneur metrics indicates the complex outcome of high QOL•Ethical. Two nations stand out in both the complex antecedent condition and complex outcome condition: Denmark and Switzerland

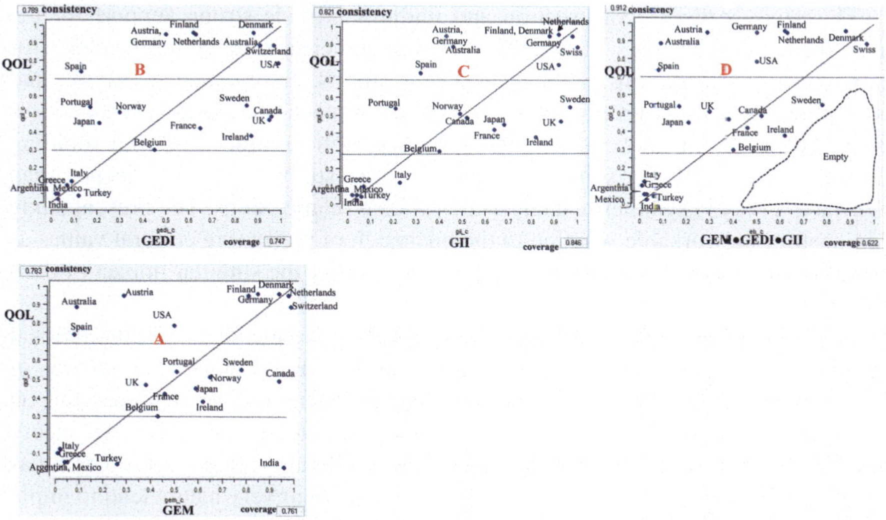

Fig. 6.14b GEM, GEDI, GII, and GEM•GEDI•GII indicating high QOL. Notes: The configura-tion model equal to GEM•GEDI•GII indicates high QOL for nations high in this configuration. The individual models for GEM, GEDI, and GII receive marginal support in indicating high QOL. The four patterns of findings provide strong support that high entrepreneurship behavior matters especially when nations commit to achieving high GEM AND high GEDI AND high GII

three nations low in GEM•GEI•GII manage to achieve high QOL—Australia, Austria, and Spain. Consequently, being high in GEM•GEI•GII assures being high in QOL, but being low in GEM•GEI•GII does not assure being low in QOL.

8 Discussion of Contributions to Theory and Limitations

8.1 Schwartz' Cultural Values Theory Is Useful for Understanding/Predicting Outcomes

Given that the definition of culture refers to the complex whole of the beliefs, values, attitudes, and behaviors of a group, society, or nation, the present study is one of the first that examines cultures' consequences on need motivations, entrepreneurship activities, quality of life, and corruption. Heretofore, empirically, the substantial majority of extant cultural value studies focus on measuring the separate impacts of each cultural values via symmetric tests such as multiple regression analysis (MRA) and structural equation models (SEM). Symmetric tests and theoretical statements fail to examine cultures as complex wholes. The present study embraces an asymmetric theoretical and empirical stance that permits advancing theory, empirical analysis, and practical recommendations regarding cultural values, need motivations, and entrepreneurship activities as complex wholes.

This study's theory construction and findings provide strong support for the usefulness of Schwartz's (1992, 2018) seven dimension of cultural values for explaining and predicting nations' need motivations, entrepreneur behavior, and corruption. In actually proposing and testing the impact of cultural values as complex wholes rather than examining the impact of each cultural value while holding the others constant via symmetric analysis, the study here goes deeper than Schwartz's (2018) symmetric analysis of variable relationships. The present study demonstrates a workable solution to the mismatch of presenting cultural values as multidimensional impacts theoretically and testing for the singular impact of each cultural value.

Fiss (2007, p. 1181) describes the mismatch eloquently: "Configurational approaches to organization are based on the fundamental premise that patterns of attributes will exhibit different features and lead to different outcomes depending on how they are arranged. But while theoretical discussions of configurational theory thus stress nonlinearity, synergistic effects, and equifinality, empirical research has so far largely drawn on econometric methods that by their very nature tend to imply linearity, additive effects, and unifinality. This mismatch has caused a number of problems. For example, the classic linear regression model treats variables as competing in explaining variation in outcomes rather than showing how variables combine to create outcomes. By focusing on the relative importance of rival variables, a correlational approach has difficulty treating cases as configurations and examining combinations of variables. This becomes particularly evident in the fact that regression analysis focuses on the unique contribution of a variable while holding constant the values of all other variables in the equation." Embracing a case-based asymmetric approach to both theory construction and data analysis offers a workable solution for overcoming this pervasive current (i.e., research during the decades before 2020) mismatch. Prior research in organizational behavior using theory and sets of data separate from the present study support this workable solution to the now pervasive asymmetric theory coupled with symmetric analysis mismatch (e.g., Fiss 2011; Frösén et al. 2016; Brenes et al. 2017; Woodside et al. 2016).

The present study offers a revision and improvement to Schwartz's (2018) view that the seven cultural value orientations fit into three bipolar cultural value dimensions, autonomy versus embeddedness, egalitarianism versus hierarchy, and harmony versus mastery, and eliminates the inherent problems (Peter et al. 1993; Teas 1993) of using difference values in data analysis. The findings include asymmetric set theory-based models that demonstrate that high autonomy AND embeddedness, egalitarianism AND hierarchy, and harmony AND mastery sometimes occur in complex antecedent conditions that accurately predict high, as well as low, outcome conditions. The same observation receives support for the configuration of low autonomy AND low embeddedness, low egalitarianism AND low hierarchy, and low harmony AND low mastery appearing within specific models indicating high scores for outcomes consistently. Conceptualizing three bipolar-culture-value dimensions is an inaccurate oversimplification for several models of how cultural value configurations impact outcome conditions.

The present study supports the perspective that theory construction and empirical research on cultural values, need motivations, entrepreneur behavior, quality of life, and corruption can advance usefully on a foundation of complexity theory and matching case-based theory and data analysis. This foundation and analytic tools make use of "statistical sameness testing" (Hubbard 2015) that avoids the "corrupt research" (Hubbard 2015; cf. Ziliak and McCloskey 2008) practices of using null hypothesis significance testing (NHST) by replacing such practices with testing for consistency (accuracy) in predicting outcomes. Woodside (2016) refers to statistical sameness testing as "somewhat precise outcome testing (SPOT)," whereby the finding for individual cases should be the same asymmetrically—high scores in the simple or complex antecedent condition should indicate high scores for the simple or complex outcome condition.

Constructing and testing directional variable hypotheses is a bad science practice from several perspectives even when the correlations among the variables are highly significant statistically. Statistically significance always occurs by increasing the number of cases in a study. Researchers can achieve statistical significance using a table of random numbers for data and stepwise regression analysis (Armstrong 2012). Researchers fail to capture the complexities in realities by not examining complex combinations of configurations of multiple "independent" variables beyond the study of two- or (rarely) three-way interactions (Ragin 1997). Cases showing relationships contrarian to hypotheses receive scant attention in research employing symmetric theory and tools. Alternatives to symmetric theory construction and analysis are available that enable examinations and theory construction of the complexities observable in facing complex realities. The seemingly small correction of changing from focusing on positive or negative symmetric directions of variable relationships to focusing on configurations of antecedent conditions predicting outcomes asymmetrically is actually a huge improvement. This case-focused improvement in theory construction and data analysis is medicine that cures McClosky's (2002) sadness:

> The progress of economic science has been seriously damaged [via using variable directional hypothesis and NHST]. You can't believe anything that comes out of the Two Sins [(i.e., being content with only qualitative variable-directional predictions in both theory and applied work]. Not a word. It is all nonsense, which future generations of economists are going to have to do all over again. Most of what appears in the best journals is unscientific rubbish. I find this unspeakably sad. (McCloskey 2002, p. 55)

8.2 *McClelland's Need Motivation Theory Is Useful for Understanding/Predicting Outcomes*

The present study supports and extends McClelland's (1961, 1965, 1987; McClelland and Winter 1969) perspectives that cultures impact national need motivations and that need motivations impact entrepreneurial behavior and quality of life. However, the findings support the view that McClelland's (1961) proposal claims too much for high nAch and entrepreneur success. Both high and low nAch occur in models that accurately predict high as well as low QOL in configurations containing entrepreneur activities (refer to the findings in Table 6.10). The findings in the present study support embracing a more complex view of how need motivations affect success than McClelland's advocacy of high nAch alone.

8.3 Entrepreneurship, Ethical Behavior, and Quality of Life

The present study findings support and extend the perspective (Davis and Ruhe 2003; Woodside et al. 2016) that nations with high entrepreneurship behavior have high QOL consistently. The findings in the present study support and extend the view that high GEM, GEDI, or GII separately is insufficient for indicating high QOL but that the occurrence of high scores for all three EB metrics does indicate high QOL. India's attempt to nurture the development of entrepreneurship by funding a wide range of GEM activities continues to fail in moving the nation toward high QOL—high GEM coupled with low GEDI and low GII does not support high QOL and indicates a low QOL outcome.

Consistently, nations achieving high ethical behavior are nations that achieve high QOL. The present study's findings support this conclusion (see Fig. 6.13), though the possibility of exceptions to this general observation supports the view that low ethical behavior can associate with high QOL (e.g., Spain). Mostly, nations lock into a low (e.g., India), medium (e.g., USA), or high (e.g., Denmark) level of ethical behavior across decades and changes in political parties and regime changes. Though achieving substantial decreases in ethical behavior is observable, the reductions appear to last for less than two decades, end with a dramatic event, and return to the previous status quo. Witness the dramatic drop in ethical behavior by large financial firms in the USA and globally that ended with the 2008 financial crisis, and the return to a medium level of ethical behavior in the USA following the US federal government financial bailouts of banks, the enactment/ implementation of the Consumer Financial Protection Bureau in 2010, and no prison terms of bank executives (Inside Job 2010)—and the return to a high level of ethical behavior in Iceland that included prison terms for 36 bankers for a collective time of 96 years (Grettisson 2018).

8.4 Religiosity, EB, QOL, and Corruption

This study indicates valuable insights follow from the inclusion of religiosity in examining the influences of alternative cultural values configurations on need motivations, EB, QOL, and corruption. National levels of religiosity often are ingredients in nations' need motivations, EB, QOL, and levels of corruption/ethical behavior. Though low religiosity does not guarantee high ethical behavior, high religiosity consistently indicates nations high in corruption. The claim is that nations' religiosity is a very useful input for the study of cultures' consequences and not that the religiosity causes corruption. The latter viewpoint fails to recognize the complexities of how culture affects national behaviors.

8.5 Limitations

The limitations of the present study include the limited number of nations covered by the examination of the nine propositions. Testing the specific models in the findings for predictive validity by including a holdout sample of nations would improve the veracity of the study's conclusions—or refute them. The study does include constructing and testing asymmetric models ex ante—before examining models constructed by running the fsQCA software program for model analysis. The findings do provide solid support for the models developed from insights and the literature. Research on cultures as complex wholes (i.e., configurations of cultural values with and without additional cultural dimensions such as religiosity) would benefit from directed comparisons of the consistencies of predictions by different theoretical models (e.g., Schwartz's versus Hofstede's theories). The study on culture, entrepreneurship, QOL, religiosity, and corruption by Woodside et al. (2016) examines cultures as complex wholes using data for Hofstede's value dimensions for 28 countries. Not examining the veracity of models in the present study on predicting entrepreneur behavior, QOL, and corruption for nations beyond the ones in the study's data set is a limitation.

The present study is not a field experiment with treatment and control conditions and random assignment of nations to treatment versus control conditions. The associations that the present study reports are not scientific estimates of cause-and-effect relationships of cultural values, need motivations, entrepreneurship, quality of life, or corruption. Field experiments of economic and social issues and the efficacy of government EB development policies are possible (Banerjee and Duflo 2009; Duflo and Kremer 2005), and the findings in the present study indicate the substantial value of field experiments of the impact of EB on increasing nations' QOL. Research on cultures as complex whole (i.e., configurations of cultural values with and without additional cultural dimensions such as religiosity) would benefit from directed comparisons of the consistencies of predictions by different theoretical models (e.g., Schwartz's versus Hofstede's theories). The study on culture, entrepreneurship, QOL, religiosity, and corruption by Woodside et al. (2016) examines cultures as complex wholes using data for Hofstede's value dimensions for 28 countries. Not examining the veracity of models in the present study on predicting entrepreneur behavior, QOL, and corruption for nations beyond the ones in the study's data set is a limitation.

9 Recommendations for National Policies and Conclusions

9.1 Revising National Cultural Value and Need Motivation Configurations

Cultural value configurations and need motivation configurations are complex wholes that change slowly, if at all, over many decades or centuries. However, change/reform programs to redirect/shift group, societal, and national cultures substantially as complex wholes do occur. The attempt in Turkey by military general and then President (1923–1938) Mustafa Kemal Atatürk in accordance with "Kemalist ideology" (i.e., sweeping political, social, cultural, and religious "reforms" designed to separate the new Turkish state from its Ottoman predecessor and embrace a Westernized way of living) is one such example of cultural change. Central to these reforms was the belief that Turkish society would have to westernize itself both politically and culturally in order to modernize. These political reforms included a number of fundamental institutional changes that brought an end to many traditions and followed a carefully planned program to unravel the complex system that had developed over the centuries. The US abolition of slavery (1865), the 1954 Brown versus Board of Education verdict, the Voting Rights Act of 1965, the Fair Housing Act of 1965, and the Civil Rights Acts of 1964 and 1968 collectively represent a national policy to end abuse and intolerance against one ethnic group by another.

Attempts starting in the late 1800s and ongoing in Norway, Denmark, New Zealand, Australia, additional Nordic nations, and later in the USA that focus on eliminating gender abuse and intolerance directed toward women by passage of reform laws and "gender mainstreaming" policies represent strategies to change cultures as complex wholes to decrease/eliminate bias and abuse against women and to nurture gender equality in employment, pay, job promotions, education, and additional life activities. Reading Scheidel's (2017) *The Great Leveler: Violence and the History of Inequality from the Stone Age to the Twenty-First Century* is useful for understanding how nations have achieved reductions in inequality and changes in cultural value configurations across one or more decades. Scheidel (2017) describes four pillars for major reductions in inequality in nations: warfare, revolution, state collapse, and plague. The assumption here is that low levels of inequality are a feature of high QOL in a nation—economic wealth is distributed widely in society. In 2018, *The Economist* (a weekly newspaper) asked Scheidel the following question: Is society incapable of tackling income inequality peacefully? Scheidel's (2018) response: "No, but history shows that there are limits. There is a big difference between maintaining existing arrangements that successfully check inequality—Scandinavia is a good example—and significantly reducing it. The latter requires real change and that is always much harder to do: think of America or Britain, not to mention Brazil, China or India. The modern welfare state does a reasonably good job of compensating for inequality before taxes and transfers. However, for more substantial levelling to occur, the established order needs to be shaken up: the greater the shock to the system, the easier it becomes to reduce privilege at the top." The central point is that some critical junctures that may span decades lead some nations to introduce change/reform policies and laws that are explicitly directed at revising their own cultural value configurations. Notice in Table 6.9, part A, that achieving a highly egalitarian nation is an ingredient in all four models for cultural value configurations indicating high QOL. However, high egalitarianism is a feature in three of the complex antecedent conditions for low QOL—see part B of Table 6.9. The impact of high egalitarianism on QOL and ~QOL involves complex configurations with additional ingredients. Egalitarianism is a necessary but insufficient condition for indicating high QOL.

How can a nation improve QOL via changes in cultural value configurations (that includes but does is not restricted to reductions in economic inequality) without resorting to warfare, revolution, state collapse, and plague? Recognizing that specific cultural value configurations do indicate QOL while others indicate ~QOL is a first step. A second step is to recognize that a few specific cultural value configurations indicate high entrepreneur activities and others indicate low entrepreneur activities (refer to Tables 6.6, 6.7, and 6.8). A third step is to recognize that a few specific entrepreneur configurations indicate high QOL while others indicate ~QOL. The fourth and fifth steps include similar observations for cultural value configurations, need motivation configurations, and QOL.

In a primary debate with Hilary Clinton, "Look to Denmark" was US Presidential candidate Bernie Sanders' comment about attempting to increase QOL by reforming US laws, policies, (increasing) governmental services, and (increasing) taxes. Clinton responded, "But we are not Denmark! I love Denmark. We're the United States of America and it's our job to rein in the excesses of capitalism so it doesn't run amok!" (Tankersley 2015). The Danes pay high taxes, but they get back some of the world's top rankings for happiness, well-being, and work-life balance—and Denmark's economic inequality (the Gini index) is low, while the USA's inequality index is high and now rising.Following the Sanders and Clinton exchange, Jackson commented (2015):

> Hillary, why boast "We are *not* Denmark" when that nation, instead of having one gun per every citizen, as we do in the United States, has nearly one bike per person? What is your point when Copenhagen's subways and trains leave America's public transit and Amtrak in the dust? What's not to like about a nation (unless you're the Koch Brothers) that has more than 500 offshore wind turbines spinning and is first in the global race to 100-percent clean energy by 2050? And given your unabashed reminders that you would be America's first woman president, don't you feel the least bit silly since Denmark's gender pay gap is half that of the US? So good for you, Hillary, to burnish your centrist credentials by dissing what your own husband said is a "pioneer" nation that inspires a "great new age of possibility." It makes for a great line for those who've never been to Denmark. For those who have, it suggests that perhaps you will not be quite the fighting pioneer in the White House than you seem to promise.

9.2 Nurturing Change in Need Motivation Configurations

The findings here do not support McClelland's perspective that high nAch is a necessary condition for nations to achieve high entrepreneurship activities or high QOL. For example, in the case example of Denmark's need motivation configuration, nations high in low nAch AND low nPow AND high nAff indicate high QOL (Table 6.8, part A). Focusing on any one of the three need motivations does not provide a useful model for indicating high QOL or ~QOL consistently. The findings do support the perspective the specific need motivation configurations indicate high QOL and ~QOL. The need motivation configuration for the USA (i.e., nAch•nAff•nPow) does indicate high GII and high QOL, but this need motivation configuration applies to only two nations (i.e., Canada and the USA).

Building on McClelland's (1961) proposition that high nAch is a principal driver for entrepreneur success, a field quasi-experiment by McClelland and Winter (1969) was an attempt to increase nAch among individuals in a small group of business executives in India. The quasi-experiment included completing learning exercises by the small group to learn how to think, talk, and perform tasks as an individual having high nAch would perform these actions. McClelland and Winter (1969) stress the need to complete their nAch training program in a group so that group members could reinforce and support all group members' increases in nAch over months and years. The before/after findings of their quasi-experiment and comparisons with an identified control group (without random assignment) failed to support their hypotheses. A review (Seibel, 1969) points out a number of telling flaws in the design of McClelland and Winter's (1969) study. In comparison to attempts to increase entrepreneurship and QOL by alternative strategies, the proposal for a nation or small groups to increase entrepreneurship activities and QOL via increasing nAch or configurations of need motivations may be too indirect to be effective.

9.3 Nurturing Entrepreneurship Activities

This study provides both theoretical and empirical support to the perspective that national government policies and reforms supporting entrepreneurship and innovation can be effective in certain configurations. The study provides strong evidence that nations supporting entrepreneurship activities are sufficient for high quality of life, given that the support is pervasive. The findings support the conclusion that substantial support of entrepreneur activities as measured by GEM is necessary but insufficient condition for high QOL. Also, the findings support the conclusion that substantial support of entrepreneur activities as measured by GEDI is necessary but insufficient condition for high QOL. Nations high in GII alone indicate high QOL (consistency = 0.82)—to some extent, achieving high GII alone is sufficient for achieving high QOL. However, the findings support the perspective that the configuration to achieving high GII is by achieving high GEM and high GEDI (consistency = 0.912). The simplified, but still complex, configuration of high GEM•GII also indicates high QOL consistently (consistency = 0.906). Ignoring GEDI scoring, five nations have high scores in GEM•GII, and all five of these nations have high QOL: Denmark, Finland, Germany, the Netherlands, and Switzerland. The complex model GEDI•GII (ignoring GEM scores) results in a model that indicates high QOL but with a lower level of consistency (consistency = 0.832).

9.4 Ethical Behavior Versus Corruption

The findings indicate the striving to end widespread corruption and maintaining/achieving high ethical behavior may be worthwhile pursuits as a national policy and in strengthening laws and regulations to achieve these objectives. Nine nations have managed to achieve high ethical standards (Fig. 6.13). None of the nine have low QOL. In comparison to these nine nations, most likely some of the laws, regulations, and policies governing ethical behavior and corruption are missing systematically among the eight nations low in ethical behavior. Given the finding that ethical behavior is highly predictive of QOL, selection and implementation of tough legal standards of marketplace behavior existing in nations high in ethical behavior by all nations are steps that may help increase QOL. This perspective is easy to write but hard to enact without the arrival of compelling national leaders and possibly only in contexts of national crises (cf. Seibel 2017).

9.5 Conclusion

Complexity theory tenets are a useful foundation for moving beyond proposing variable directional hypotheses to construct theory and test complex configurational antecedent conditions that accurately indicate simple and complex outcomes. Certain causal statements of complex antecedent conditions of cultural values do indicate high entrepreneur behavior and QOL. These conditions differ in content from complex antecedent conditions indicating nations low in entrepreneur behavior and negation in QOL. High entrepreneurship activity across any two and all three dimensions (GEM, GEDI, and GII) indicates nations high in QOL. Nations are identifiable as having high levels of all three entrepreneur actions; these nations are all high in QOL and low in unethical behavior. India's high support of GEM activities but not GEDI nor GII is suggestive of what will not work to increase a nation's QOL. Surprisingly, the negation of religiosity indicates nations high in ethical behavior. High ethical behavior indicates nations high in QOL. Asymmetric modeling permits the matching of case-based theory construction with case-based data analysis—a substantial advance for theory construction and practical improvements in national entrepreneurial policies.

Appendix A: Schwartz's Cultural Values Data

Abbrev	Country	Harmony	Embedded	Hierarchy	Mastery	Affautonomy	Intautonomy	Egalitarian
ARG	Argentina	3.98	3.52	2.10	3.92	3.73	4.34	4.96
AUL	Australia	3.99	3.59	2.29	3.97	3.86	4.35	4.79
AUS	Austria	4.31	3.11	1.75	3.92	4.29	4.90	4.89
BEL	Belgium	4.35	3.25	1.69	3.84	3.94	4.64	5.20
CAN	Canada	3.83	3.46	2.09	4.12	4.00	4.50	4.80
DEN	Denmark	4.16	3.19	1.86	3.91	4.30	4.77	5.03
FIN	Finland	4.34	3.37	1.80	3.66	3.96	4.93	4.90
FRA	France	4.21	3.20	2.21	3.72	4.39	5.13	5.05
GER	Germany	4.62	3.03	1.87	3.86	4.11	4.99	5.07
GRB	Great Britain, UK	3.91	3.34	2.33	4.01	4.26	4.62	4.92
GRE	Greece	4.40	3.41	1.83	4.25	3.92	4.39	4.84
IND	India	3.92	3.97	3.05	4.28	3.48	4.02	4.45
IRE	Ireland	3.77	3.41	2.09	4.04	4.05	4.54	4.90
ITA	Italy	4.62	3.46	1.60	3.81	3.30	4.91	5.27
JPN	Japan	4.21	3.49	2.65	4.06	3.76	4.78	4.36
MEX	Mexico	4.50	3.90	2.13	3.90	2.83	4.36	4.73
NET	Netherlands	4.05	3.19	1.91	3.97	4.13	4.85	5.03
NOR	Norway	4.40	3.45	1.49	3.85	3.69	4.68	5.12
POR	Portugal	4.27	3.43	1.89	4.11	3.62	4.53	5.21
SPA	Spain	4.47	3.31	1.84	3.80	3.67	4.99	5.23
SWE	Sweden	4.46	3.12	1.83	3.81	4.24	5.09	4.90
SWI	Switzerland	3.94	3.34	2.42	3.97	4.24	4.66	4.92
TUR	Turkey	4.23	3.77	2.97	3.98	3.37	4.45	4.77
USA	USA	3.46	3.67	2.37	4.09	3.87	4.19	4.68

Appendix B: National Data for Religiosity; McClelland's nAch, nAff, and nPow; GII, GEM, and GEDI; GDPppp; Quality of Life; and Transparency (Negation of Corruption)

Abbrev	Religiosity	needach	needaff	needpow	gii18	gemmed	gedi	gdppppc17	qualitylife17	Transparency
ARG	66	3.38	1.29	0.95	30.7	23.9	24	20,677	119.71	39
AUL	32	2.38	1.24	0.52	52	34.7	75.5	49,882	172.4	77
AUS	55	1.86	1.14	0.43	51.3	52.7	66	49,247	178.53	75
BEL	33	0.43	1.52	0.95	50.5	60.4	63.7	46,301	147.51	75
CAN	42	2.29	2.19	1.24	53	82.9	79.2	48,141	158.19	82
DEN	18	1.05	1.76	0.86	59.6	82.1	74.3	49,613	180.14	88
FIN	28	1.52	1.71	1.05	59.6	75.9	67.9	44,050	179.4	85
FRA	29.5	2.38	1.19	0.48	54.5	62.1	68.5	43,550	154.43	70
GER	40.5	2.14	1.29	1.33	58.4	73.7	65.9	50,206	178.92	81
GRB	26.5	1.67	2.54	1.1	60.1	58	77.8	43,620	156.8	82
GRE	71.5	2.29	2.14	0.81	38.9	8.9	37.1	27,776	131.1	48
IND	79	2.71	1.24	1.24	35.2	86	28.4	7174	110.39	40
IRE	53.5	2.29	1.38	0.9	57.2	67.4	73.7	72,632	152.11	74
ITA	71.5	1.33	2	0.95	46.3	12.6	41.4	37,970	132.59	50
JPN	23.5	1.29	2.19	0.95	55	66.7	51.5	42,659	155.8	73
MEX	72	1.57	1.62	1.19	35.3	27.7	26.4	19,480	120.73	29
NET	24.5	1.48	2.19	0.71	63.3	91.6	68.1	53,582	177.8	82
NOR	20.5	1.71	1.95	0.57	52.6	68.3	56.6	70,590	158.85	85
POR	71.5	2.1	1.67	0.48	45.7	64.4	48.8	30,258	159.59	63
SPA	49.5	2.33	0.62	1.86	48.7	32.4	45.3	38,171	165.67	57
SWE	16.5	1.62	1.48	0.76	63.1	72.6	73.1	51,264	159.9	84

Abbrev	Religiosity	needach	needaff	needpow	gii18	gemmed	gedi	gdppppc17	qualitylife17	Transparency
SWI	41.5	1.71	1.52	0.52	68.4	95	80.4	61,360	172.86	85
TUR	89.5	3.62	0.9	0.62	37.4	51.1	44.5	26,453	118.34	40
USA	65	2.24	2.05	1.38	59.8	63.9	83.6	59,495	167.5	75

Appendix C: Configurations of Need Motivation Indicating Nations High in Ethical Behavior Versus High in Corruption

(A) **Predicting nations high in ethical behavior by needs motivation: model: n_corrupt_c = f(pow_cc, aff_cc, ach_cc)**

INTERMEDIATE SOLUTION: frequency cutoff: 1.00; consistency cutoff: 0.85.

Models	Raw coverage	Unique coverage	Consistency
1. ~aff•~ach	0.36	0.05	0.86
2. ~powcc•~ach_cc	0.52	0.21	0.90

Solution coverage: 0.57; solution consistency: 0.86

(B) **Predicting nations high in corruption using needs motivating: model: corrupt = f(pow_cc, aff_cc, ach_cc)**

INTERMEDIATE SOLUTION: frequency cutoff: 1.00; consistency cutoff: 0.81.

Models	Raw coverage	Unique coverage	Consistency
1. pow•~aff•ach	0.34	0.23	0.83
2. ~pow•aff•ach	0.23	0.13	0.81

Solution coverage: 0.46; solution consistency: 0.81.

Appendix D: Schwartz's Cultural Values and Religiosity Indicating High (Low) nAch

(A) **Indicating high nAch: ach_cc = f(relig_c, egalitar_c, intel_auto_c, aff_auto_c, mast_c, hier_c, embed_c, harm_c)**

INTERMEDIATE SOLUTION: frequency cutoff: 1.00; consistency cutoff: 0.92

Models	Raw coverage	Unique coverage	Consistency
1. relig_c•egalitar_c•intel_auto_c•aff_auto_c•~hier_c•~embed_c•~harm_c	0.26	0.03	0.96
2. relig_c•egalitar_c•intel_auto_c•aff_auto_c•mast_c•~hier_c•~embed_c	0.38	0.15	0.94
3. relig_c•egalitar_c•~intel_auto_c•aff_auto_c•mast_c•hier_c•~embed_c•~harm_c	0.13	0.02	0.99
4. relig_c•~egalitar_c•~intel_auto_c•aff_auto_c•mast_c•hier_c•embed_c•~harm_c	0.11	0.04	1.00
5. relig_c•egalitar_c•intel_auto_c•~aff_auto_c•mast_c•hier_c•~embed_c•harm_c	0.11	0.05	0.94

Solution coverage: 0.53; solution consistency: 0.95

(continued)

(B) **Indicating the negation of nAch: not_ach_cc = f(relig_c, egalitar_c, intel_auto_c, aff_auto_c, mast_c, hier_c, embed_c, harm_c)**			
INTERMEDIATE SOLUTION: frequency cutoff: 1.00; consistency cutoff: 0.86			
Models	**Raw coverage**	**Unique coverage**	**Consistency**
1. relig_c•egalitar_c•intel_auto_c•~aff_auto_c•~mast_c•~hier_c•harm_c	0.15	0.12	0.88
2. ~relig_c•egalitar_c•intel_auto_c•aff_auto_c•mast_c•~hier_c•~embed_c•harm_c	0.22	0.16	0.87
3. ~relig_c•~egalitar_c•intel_auto_c•aff_auto_c•mast_c•hier_c•~embed_c•harm_c	0.12	0.07	0.91
Solution coverage: 0.41; solution consistency: 0.89.			

Appendix E: The Findings Support P3 for nAff: Religiosity Embeds in Cultural Values Configurations Indicating nAff and the Negation of nAff

(A) **Model: aff_cc = f(relig_c, egalitar_c, intel_auto_c, aff_auto_c, mast_c, hier_c, embed_c, harm_c)**			
INTERMEDIATE SOLUTION: frequency cutoff: 1.00; consistency cutoff: 0.83.			
Models	**Raw coverage**	**Unique coverage**	**Consistency**
1. relig_c•egalitar_c•intel_auto_c•~aff_auto_c•~mast_c•~hier_c•harm_c	0.15	0.12	0.84
2. ~relig_c•egalitar_c•intel_auto_c•aff_auto_c•mast_c•~hier_c•~embed_c•~harm_c	0.22	0.14	0.83
3. relig_c•egalitar_c•~intel_auto_c•aff_auto_c•mast_c•hier_c•~embed_c•~harm_c	0.12	0.04	0.84
4. ~relig_c•~egalitar_c•intel_auto_c•aff_auto_c•mast_c•hier_c•~embed_c•harm_c	0.12	0.08	0.92
Solution coverage: 0.46; solution consistency: 0.85.			

(B) **Model: not_aff_cc = f(relig_c, egalitar_c, intel_auto_c, aff_auto_c, mast_c, hier_c, embed_c, harm_c)**			
INTERMEDIATE SOLUTION: frequency cutoff: 1.00; consistency cutoff: 0.83			
Models	**Raw coverage**	**Unique coverage**	**Consistency**
1. relig_c•egalitar_c•intel_auto_c•aff_auto_c•~mast_c•~hier_c•~embed_c	0.31	0.24	0.94
2. relig_c•~egalitar_c•~intel_auto_c•aff_auto_c•mast_c•hier_c•embed_c•~harm_c	0.09	0.04	0.85
3. relig_c•egalitar_c•intel_auto_c•~aff_auto_c•~mast_c•~hier_c•embed_c•harm_c	0.07	0.02	0.83
4. relig_c•egalitar_c•intel_auto_c•~aff_auto_c•mast_c•hier_c•~embed_c•harm_c	0.11	0.04	1.00
Solution coverage: 0.44; solution consistency: 0.90.			

Appendix F: The Findings Support P3 for nPow: Religiosity Embeds in Cultural Values Configurations for nPow and the Negation of nPow

(A) **Model: nPow_c = f(relig_c, egalitar_c, intel_auto_c, aff_auto_c, mast_c, hier_c, embed_c, harm_c)**			
INTERMEDIATE SOLUTION: frequency cutoff: 1.00; consistency cutoff: 0.87			
Models	**Raw coverage**	**Unique coverage**	**Consistency**
1. relig_c•egalitar_c•intel_auto_c•~aff_auto_c•~mast_c•~hier_c•harm_c	0.16	0.13	0.90
2. relig_c•egalitar_c•~intel_auto_c•aff_auto_c•mast_c•hier_c•~embed_c•~harm_c	0.13	0.05	0.90
3. relig_c•~egalitar_c•~intel_auto_c•aff_auto_c•mast_c•hier_c•embed_c•~harm_c	0.11	0.04	0.91
Solution coverage: 0.30; solution consistency: 0.91.			
(B) **Model: not_pow_cc = f(relig_c, egalitar_c, intel_auto_c, aff_auto_c, mast_c, hier_c, embed_c, harm_c)**			
INTERMEDIATE SOLUTION: frequency cutoff: 1.00; consistency cutoff: 0.88.			
Models	**Raw coverage**	**Unique coverage**	**Consistency**
1. egalitar_c•intel_auto_c•aff_auto_c•mast_c•~hier_c•~embed_c•harm_c	0.36	0.25	0.94
2. relig_c•egalitar_c•intel_auto_c•aff_auto_c•~mast_c•~hier_c•~embed_c•~harm_c	0.12	0.02	0.92
3. relig_c•egalitar_c•intel_auto_c•~aff_auto_c•mast_c•hier_c•~embed_c•harm_c	0.10	0.06	0.89
Solution coverage: 0.44; solution consistency: 0.91.			

References

Ács, Z. J., Szerb, L., & Autio, E. (2015). *Global entrepreneurship and development index powered by GEDI*. Washington, DC: The Global Entrepreneurship and Development Institute.

Anderson, J. L. (2018). Sick of corruption and Trump: Voters embrace the maverick leftist Andrés Manuel López Obrador. The New Yorker, June 18, downloaded on October 20, 2019 at: https://www.newyorker.com/magazine/2018/06/25/a-new-revolution-in-mexico

Armstrong, J. S. (2012). Illusions in regression analysis. *International Journal of Forecasting, 28*, 689–694.

Banerjee, A. V., & Duflo, E. (2009). The experimental approach to development economics. *Annual Review of Economics, 1*, 151–178.

Bardi, A., & Schwartz, S. H. (2003). Values and behavior: Strength and structure of relations. *Personality and Social Psychology Bulletin, 29*, 1207–1220.

Bergmann, H., Mueller, S., & Schrettle, T. (2013). The use of global entrepreneurship monitor data in academic research: A critical inventory and future potentials. *International Journal of Entrepreneurial Venturing, 6*, 242–276.

Brenes, E. R., Ciravegna, L., & Woodside, A. G. (2017). Constructing useful models of firms' heterogeneities in implemented strategies and performance outcomes. *Industrial Marketing Management, 62*, 17–35.
Chelariu, C., Brashear, T. G., Osmonbekov, T., & Zait, A. (2008). Entrepreneurial propensity in a transition economy: Exploring micro-level and meso-level cultural antecedents. *Journal of Business & Industrial Marketing, 23*(6), 405–415.
Davis, J. H., & Ruhe, J. A. (2003). Perceptions of country corruption: Antecedents and outcomes. *Journal of Business Ethics, 43*, 275–288.
Dodd, D. S., & Gotsis, G. (2007). The interrelationships between entrepreneurship and religion. *Entrepreneurship and Innovation, 8*, 93–104.
Duflo, E., & Kremer, M. (2005). Use of randomization in the evaluation of development effectiveness. *Evaluating Development Effectiveness, 7*, 205–231.
Dutta, S., Lavnin, B., & Wunsch-Vincent, S. (Eds.). (2015). The Global Innovation Index 2015: Effective innovation policies for development (Fontainebleau, Ithaca, and Geneva. Downloaded on October 26, 2016 at http://www.wipo.int/edocs/pubdocs/ en/wipo_gii_2015.pdf).
Fayolle, A., Liñán, F., & Moriano, J. A. (2014). Beyond entrepreneurial intentions: Values and motivations in entrepreneurship. *International Entrepreneurship Management Journal, 10*, 679–689.
Ferriss, A. L. (2000). The quality-of-life among U.S. states. *Social Indicators Research, 49*, 1–23.
Ferriss, A. L. (2002). Religion and the quality-of-life. *Journal of Happiness Studies, 3*, 199–215.
Finke, R., & Adamcyzyk, A. (2008). Cross-national moral beliefs: The influence of national religious context. *Social Quarterly, 49*, 617–652.
Fiss, P. C. (2007). A set-theoretic approach to organizational configurations. *Academy of Management Review, 32*, 1180–1198.
Fiss, P. C. (2011). Building better casual theories: A fuzzy set approach to typologies in organizational research. *Academy of Management Journal, 54*, 393–420.
Fontaine, J. R. J., Poortinga, Y. H., Delbeke, L., & Schwartz, S. H. (2008). Structural equivalence of the values domain across cultures: Separating sampling fluctuations from meaningful variation. *Journal of Cross-Cultural Psychology, 39*, 345–365.
Frösén, J., Luoma, J., Jaakkola, M., Tikkanen, H., & Aspara, J. (2016). What counts versus what can be counted: The complex interplay of market orientation and marketing performance measurement. *Journal of Marketing, 80*, 60–78.
Global Index of Religiosity and Atheism. (2012).WIN-Gallup International GLOBAL INDEX OF RELIGIOSITY AND ATHEISM: Zurich. http://www.wingia.com/web/files/news/14/file/14.pdf. Downloaded on August 15, 2018.
Grettisson, V. (2018). 36 bankers, 96 years in jail. https://grapevine.is/news/2018/02/07/36-bankers-96-years-in-jail/. Downloaded by A. G. Woodside, September 10, 2018.
Harrison, G. V., & List, J. (2004). Field experiments. *Journal of Economic Literature, 42*, 1009–1055.
Harvard Divinity School. (2018). https://rlp.hds.harvard.edu/religious-influences-are-embedded-cultures. Downloaded on August 19, 2018.
Hofstede, G. (2001). *Culture's consequences: Comparing values, behaviors, institutions, and organizations across nations* (2nd ed.). Thousand Oaks, CA: Sage.
Hubbard, R. (2015). *Corrupt research: The case for reconceptualizing empirical management and social science*. Los Angeles: Sage.
Inglehart, R., & Baker, W. E. (2000). Modernization, cultural change, and the persistence of traditional values. *American Sociological Review, 65*, 19–51.
Inside Job. (2010). https://www.theguardian.com/film/2011/feb/17/inside-job-financial-crisis-bankers-verdicts. Downloaded by A.G. Woodside, September 10, 2018.
Jackson, D. Z. (2015). Memo to Hillary Clinton: What's not to like about Denmark? https://www.bostonglobe.com/opinion/2015/10/19/memo-hillary-clinton-what-not-like-about-denmark/PoV2OD3KpysMFCAz2XLI0K/story.html. Downloaded by A. G. Woodside, September 12, 2018.

Jin, B., & Cho, H. J. (2018). Examining the role of international entrepreneurial orientation, domestic market competition, and technological and marketing capabilities on SME's export performance. *Journal of Business & Industrial Marketing, 33*(5), 585–598.

Liñán, F., Fernández, J., & Romero, I. (2013). Necessity and opportunity entrepreneurship: The mediating effect of culture. *Revista de Economía Mundial, 33*, 21–47.

McClelland, D. C. (1961/1976). *The achieving society*. Oxford, UK: Irvington.

McClelland, D. C. (1965). N achievement and entrepreneurship: A longitudinal study. *Journal of Personality and Social Psychology, 1*(4), 389–392.

McClelland, D. C. (1987). Characteristics of successful entrepreneurs. *Journal of Creative Behavior, 21*, 219–233.

McClelland, D. C., & Winter, D. G. (1969). *Motivating economic achievement: Accelerating economic development through psychological training*. New York: Free Press.

McClelland, D. C., Atkinson, J. W., Clark, R. A., & Lowell, E. L. (1953). *Century psychology series. The achievement motive*. East Norwalk, CT: Appleton-Century-Crofts.

McCloskey, D. (2002). *The secret sins of economics*. Chicago, IL: Prickly Paradigm Press.

Miron, D., & McClelland, D. C. (1979). The impact of achievement motivation training on small businesses. *California Management Review., 21*, 13.

Murray, H. A. (1938). *Explorations in personality*. New York: Oxford University Press.

Okulicz-Kozaryn, A. (2010). Religiosity and life satisfaction (A multilevel investigation across nations). *Mental Health, Religion & Culture, 13*(2), 155–169.

Omeje, K. (Ed.). (2008). *Extractive economies and conflicts in the global south: Multi-regional perspectives on rentier politics*. Aldershot: Ashgate.

Peter, J. P., Churchill, G. A., & Brown, T. J. (1993). Caution in the use of difference scores in consumer research. *Journal of Consumer Research, 19*, 655–662.

Ragin, C. C. (1997). Turning the tables: How case-oriented research challenges variable-oriented research. *Comparative Social Research, 16*, 27–42.

Ragin, C. C. (2008). *Redesigning social inquiry: Fuzzy sets and beyond*. Chicago: University of Chicago Press.

Rumelt, R. (2011). *Good strategy, bad strategy*. London: Profile Books.

Schacter, D., Gilbert, D., Wegner, D., & Hood, B. (2011). *Psychology: European edition*. Basingstoke: Palgrave Macmillan.

Scheidel, W. (2017). *The great leveler: Violence and the history of inequality from the stone age to the twenty-first century*. Princeton, NJ: Princeton University Press.

Scheidel, W. (2018). The great leveler: Violence and the history of inequality from the Stone Age to the Twenty-First Century. Princeton: Princeton University Press.

Schuetze, C. F. (2018). Denmark's model: Quality living, with benefits. https://www.usnews.com/news/best-countries/articles/2018-01-25/denmarks-social-policies-generally-remain-a-model-say-experts. Downloaded September 5, 2018.

Schumpeter, J. A. (1934). *The theory of economic development*. (Translated by R. Opie). Cambridge, MA: Harvard University Press.

Schumpeter, J. A. (1942). *Capitalism, socialism and democracy*. New York, NY: Harper and Row.

Schwartz, S. H. (1992). Universals in the content and structure of values: Theory and empirical tests in 20 countries. In M. Zanna (Ed.), *Advances in experimental social psychology* (Vol. 25, pp. 1–65). New York: Academic Press.

Schwartz, S. H. (2006). A theory of cultural value orientations: Explication and applications. *Comparative Sociology, 5*, 137–182.

Schwartz, S. H. (2018). National culture as value orientations: Consequences of value differences and cultural distance. Chapter forthcoming in V. Ginsburgh & D. Throsby (Eds.), *Handbook of the economics of art and culture* (Vol. 2). London: Elsevier/North Holland.

Schwartz, S. H., & Boehnke, K. (2004). Evaluating the structure of human values with confirmatory factor analysis. *Journal of Research in Personality, 38*, 230–255.

Seibel, H.D. (1969). [Book review.] Motivating economic achievement by D. C. McClelland and D. G. Winter. New York: Free Press, 1969, 409 pages; Social Forces, 48(2), 277–278.

Seibel, W. (2017). The Econmic Union, Ukraine, and the unstable east. Dinan, D. Nugent, N., Paterson, W.E., (2017) (Eds.), The European Union in crisis. New York: Macmillan International Higher Eduction, 269–293.

Singer, S., Amorós, J. E., & Arreloa, D. M. (2014). Global entrepreneurship monitor 2014 global report. New York: Global Entrepreneurship Research Association (GERA). Complete report downloaded, August 12, 2018 at http://www.babson.edu/Academics/centers/blank-center/global-research/gem/Documents/GEM%202014%20Global%20Report.pdf.

Singh, R., & Seshadri, D. V. R. (2012). Tiger, tiger, burning bright! Will India's tiger economy stimulate its business marketing space? *Journal of Business & Industrial Marketing, 27*(3), 255–260.

Stavrova O, Fletchenhauer, D., & Schlösser, T. (2013). Why are religious people happy? The effect of the social norm of religiosity across countries. Social Science Research, 42, 90–105.

Steenkamp, J.-B. E. M. (2001). The role of national culture in international marketing research. *International Marketing Review, 18*, 30–44.

Tankersley, J. (2015). Why Bernie Sanders loves Denmark but Hillary Clinton doesn't. https://www.washingtonpost.com/news/wonk/wp/2015/10/13/why-bernie-sanders-loves-denmark-but-hillary-clinton-doesnt/?noredirect=on&utm_term=.8b57fcdf329d. Downloaded by A.G. Woodside, September 12, 2018.

Teas, R. K. (1993). Expectations, performance evaluation, and consumers' perception of quality. *Journal of Marketing, 57*, 18–34.

The Economist. (2018). The lessons of violence and inequality through the ages. Interview with Walter Scheidel. https://www.economist.com/books-and-arts/2017/03/02/the-lessons-of-violence-and-inequality-through-the-ages. Downloaded by A.G. Woodside, September 13, 2018.

Transparency International. (2015). *Corruption perceptions index 2015*. Transparency International.

Urry, J. (2005). The complexity turn. *Theory, Culture & Society, 22*, 1–14.

Weick, K. (1987). Organizational culture as a source of high reliability, California Management Review, 29, 112–127.

Woodside, A. G. (2016). Diffusion and adoption of good science: Overcoming the dominant logic of NHST and the reporting of rubbish. *Journal of Business-to-Business Marketing, 23*, 327–333.

Woodside, A. G., Mir-Bernal, P., & Coduras, A. (2016). The general theory of culture, entrepreneurship, innovation, and quality-of-life: Comparing nurturing versus thwarting enterprise start-ups in BRIC, Denmark, Germany, and the United States. *Industrial Marketing Management, 53*, 136–159.

Woodside, A. G., Nagy, G., & Megehee, C. M. (2018). Four-corner outcomes in strategic management: Successful and unsuccessful paddling down versus upstream. In A. G. Woodside (Ed.), *Improving the marriage of modeling and theory for accurate forecasts of outcomes. Advances in Business Marketing & Purchasing* (Vol. 25, pp. 19–62).

Zadeh, L. (1996). Fuzzy logic: Computing with words. *IEEE Transactions on Fuzzy Systems, 4*, 103–111.

Ziliak, S. T., & McCloskey, D. N. (2008). *The cult of statistical significance: How the standard error costs us jobs, justice, and lives*. Ann Arbor: University of Michigan Press.

Author Index

A. G. Woodside (ed.), *Accurate Case Outcome Modeling*,
https://doi.org/10.1007/978-3-030-26818-3

Subject Index

A. G. Woodside (ed.), *Accurate Case Outcome Modeling*,
https://doi.org/10.1007/978-3-030-26818-3

The manufacturer's authorised representative in the EU is Springer Nature Customer Service Centre GmbH, Europaplatz 3, 69115 Heidelberg, Germany. If you have any concerns regarding our products, please contact ProductSafety@springernature.com

Printed and bound by CPI Group (UK) Ltd, Croydon, CR0 4YY
30/09/2026
02225508-0007